ASTROLOGY

history and purpose

Jane Ahlquist

unicorn house

ASTROLOGY: history and purpose

Published by

Unicorn House
P.O. Box 10,
Majors Creek N.S.W. 2622
AUSTRALIA
www.unicornhouse.org

National Library of Australia Cataloguing-in-Publication entry

Creator: Ahlquist, Jane E., 1951- author

Title: Astrology : history and purpose : a rumination on astrology, religion and Mother Earth
Jane Ahlquist
ISBN: 9780994369314
Subjects: Astrology—History
Dewey Number: 133.509

Cover design: Judit Kovacs
Illustrator: Christine Payne
Paintings: Jane Ahlquist
Layout: Michelle Lovi, Odyssey Books
Printer: IngramSpark Inc.

Note: for privacy reasons, the names on the astrology wheels have been changed.

To the Angel of Uranus —
you have been the best of friends

and to my comrade Lloyd,
a true son of the planet

CONTENTS

Introduction   1

PART ONE

1. The Glory and her Wheel   9
**- In which Astrology is found to contain clues to an old way
of initiation.**
*The Sap of the Sacrificial God; Moments of Glory; Geo and
Helio; Our Lost Capacity – the necessary Astrology; Our Lost
Capacity - the necessary Angel Body; Culture, History, Exile.*

2. Of Sabians and Stone Astronomers   19
**- In which the state of mind that would become Astrology is
located in the ancient world.**
*Anatolia; Stone; Metal; The Chamber of Venus; The Temple of
Rejoicing.*

3. Traces of Wisdom   37
**- In which the zodiac is completed and the living Spirit
departs from one branch of religion, only to make her home
in another during a critical time.**
*Meta-History; Visionary Mind; The Rise of the Bureaucrats;
Saving Enoch; Removing the Traces.*

4. The Fallen Angels   47
**- In which Astrology is brought to earth by the Fallen
Angels and enters the mythology of the age.**
*Original Myth; Revolving Stars; Prophecy to Prediction;
Corruption or Ferment?*

5. The Angel Body   **57**
- In which the history of Astrology passes inwards and enters the arena of the human soul.
The Shifting of Blame; Communities of Light; The Angel Body; The Eye of the Angel.

6. The Knot of Christos   **67**
- In which Astrology is recruited by the Angel Body, and the sheaths of the Angel Body are explained.
Rome and Jerusalem; The Heart Body; The Mission Body; The Praise Body; Eternal Rivalries; Gnostic Ripples.

PART TWO

7. The Hermetic Harmonic   **83**
- In which Hermeticism flows into sacred history and divine confusion besets the East-West border.
Knowledge Abroad; Conversations Along The Road; Poimandres, the Shepherd of Men; Tributaries.

8. The Magical Forge   **91**
- In which sacred history travels into the West across time – and waits.
Lines of Sacred History; The Monks Re-open the Books; The Jewish Heresy; A Note on Pythagoras; Astrological Stages; Hermetica Ahoy!; Unintended Consequences.

9. A Question of Witches   **107**
- Being an astrological analysis of one occult battle during the Renaissance.
Poisoned Arrows; Jumping at Shadows; Saturn, Pluto and the Wombs of Women; Nuts.

10. A Blind Date   117
- In which the scientific age opens, Astrology retreats and the other magic arises.
Blind History; First Light; The Rings of Saturn; The Great Man Fallacy; Gutting the Glory, Preserving the Magic; Thoth Incarnates.

PART THREE

11. The Other Science   129
- A sampling of prophetic stories as a means of reclaiming science.
The Astrologer's Mind; Kepler's Eye; Jupiter the Storyteller; Dream of the Red Spot; 36 Plus One; Orbital Notes; Chiron and the Contract of Humanity; Twinkle Twinkle.

12. Reading the Map   147
- An outline of various approaches to interpreting the astrology chart.
Drama or Karma? Staying With the Client; The Map as Atmosphere; The Map as Sleep;
The Map as Temple; Ancient Planets, Modern Trinity.

13. Uranus the Stranger   163
- Being a re-arrangement of mind.
The Planet; Uranus and Presence; Shock or Revelation? Uranus and Paradox; Celestial Physics; Higher Octave Application: using thought; Uranus Asleep – the ultimate paradox; The Astonished Mind; Becoming a Stranger.

14. Pluto the Vehicle   **181**
- Being a look at character itself.
The Planet; Incubation, Compression, Prophecy; The Transmission of Panic; The Fertile Regions; Pluto in Hell; Transforming Projection; Crisis, What Crisis? Resurrection and the Duty of the Knight.

15. Neptune the Glory   **199**
- Being the heart of the matter.
The Planet; Imagination or imaginings? A Forest of Fakes; Betraying the Silent Elephant; The Transits of Neptune; Mary Magdalene: a study; The Indestructible Web.

16. The Gift   **219**
- An approach to spiritual awareness in the light of earth science.
Lesser and Greater; Threshold Moments; The Doorway of Handicap; The Perfumed Land; Giftedness.

17. In trust   **233**
- Being a few final thoughts.

Bibliography   **237**

Further Acknowledgements   **243**

About the Author   **244**

INTRODUCTION

There are gates of the pathways of Night and Day,
held fast in place between the lintel above and a threshold of stone;
and they reach up into the heavens, filled with gigantic doors.

– Parmenides (b. 520 BC)

A practice can be undertaken for years before a person recognises what he or she has been really doing. So it is that I practised astrology for 35 years (with breaks and those enforced holidays when nobody comes for a while), until I knew its value in my heart.

At that time I had (and still have) two concerns. One is the restoration of the spiritual sight that is able to 'see' the immanent, feminine Glory and so discloses once again the sacred dimension of Mother Nature at a critical time. The other is the bringing of astrology out of the shadows of mockery and into the light of recognition as an authentic spiritual practice. Gradually, I began to see that these concerns are connected. For astrological practice leads to a direct perception of the inner plane energies that inhabit place and events. Our language is a hologram based on time, function and story: it is an ancient way of thinking.

Every long-time astrologer knows that there comes a moment of illumination in the practice, a breakthrough, when some additional energy is made available to the mind. At this moment, something within jumps for joy and claps its hands. It is the beginning of direct perception – the language of the heart. After that, other branches of knowledge are approached in a manner that also gently invokes the 'heart' of the matter or, rather, that seeks after a language that may invite the heart of the subject to speak. We speak of a planet in the hope that its Angel may speak to us; and we speak to the Glory in nature.

This book is for astrologers – with love – and for those beginning on the path. Our journey will take us from the historical seeds of Western astrology – the line of Sabianism and the emigrations from the network of megalithic sites – to the final public fall of astrology in the period of the rise of the Renaissance magicians and their direct descendants, the contemporary scientists.

After this, we will look at the Other Science, the science that is inseparable from storytelling, before we arrive at some practical astrology in chapters on Uranus, Neptune and Pluto – as Stranger, Glory and Vehicle – through which I hope to return astrology to its position as a powerhouse for genuine spiritual seekers.

In this too brief journey, it is above all the passage of the Glory that is sought, where she may have touched and emboldened history, both through those individuals who have carried what is called the perennial wisdom and through historical ruptures in which her influence is apparent, if suppressed. It is a book written from inside the practice of astrology and from the point of view of its gifts. The aim is to give a 'taste' for our history in the hope that astrologers may recover enough validation to deepen their search, armed with some clues.

The sampling of the Glory is a big subject to be distilled into such a small offering. Therefore, some rather brutal surgery has had to be made when choosing the historical subjects. Below are my excuses and an outline of what has been omitted.

Problems of Selection

A number of apologies must be made regarding the westernised selection that constitutes the narrative of Astrology and the Angel Body. For the nature of the story is 'Eurocentric'. As an astrologer trained in Australia, I have inherited Western astrology, with its wheel (albeit arising from the Babylonian zodiac), its Greek myths, its German psychology and its British research models. Therefore I have deemed it best to open out this history, with my own emphasis, rather than include the equally rich crossover areas of, say, the Arabian astrology with its parts and the entirely separate Chinese imagery.

However, Eurocentrism always rests upon an assumption and that assumption is that it is in the West that most of the key scientific discoveries informing later times (and which aided astrological calculation) were made and that it is the Greeks that dominated earlier knowledge and formed the basis of the Western inheritance. This is simply not true and it is our blind spot.

Take astronomy – not separate from astrology in the ancient mind – without which there can be no accurate horoscopy nor anticipation of our planetary seasons.

The West struggled with astronomy until the late Middle Ages, when the scholarship of the Arabs finally reached the new universities, entering via Spain and France. Up until that time, Western scholars could not agree on a consistent astronomy giving, say, the time of the rising of a sign of the zodiac or the accurate position of a planet at midnight (and when was midnight anyway?). Therefore, there could not have been *any* accurate horoscopes in the West before the twelfth century AD, and even in the sixteenth century the English court astrologer John Dee continued to complain about the poor mathematics behind the art.

The Arabian scholarship had its roots in turn in Egypt and Mesopotamia. Even the Greeks, whom the West wrongly consider as the foundation of its own learning, always acknowledged their debt to Egypt. In fact the great Pythagoras himself travelled widely in Mesopotamia and spent twenty years in Egypt, and the famous Pythagorean theorem was already present in these and other parts of the world. It is now known that there was a developed mathematics in India, China and Pre-Columbian America, in some cases stretching back as far as 2000 BC[1]

At the time that the West has labelled the beginning of the Dark Ages (fourth century BC), Alexandria was being established, and

---

1   The ancient Babylonians had methods for solving equations that were not improved upon until the 16th century AD

into Alexandria flowed the mathematics and scholarship of Babylon and Egypt where they met the abstract geometry of the Greeks. Right through the Dark Ages, the Arabs travelled widely in China, India and the Hellenistic world and by the sixth century AD they were synthesizing their discoveries.[2] This knowledge was passed through Baghdad and Cairo and it was from there, via the Islamic conquest of Spain, which shifted the scholarly centre of gravity to Cordoba in the 8th and 9th centuries, that mathematics and astronomy finally reached a West that had been dozing for centuries.

Yet 19th century Western historians would claim that the site of all valuable scientific objectivity lay with England, Italy and the lands between. So much for Eurocentric tendencies.

By the time these 19th century histories of science were being compiled, astrology of course had been discredited in the public domain and so a further distortion, the failure to report the impact of an integrated astrology on the mathematical thinking of civilization (particularly on the Indian subcontinent which gave us computation) occurred. Even in 20th century India, mathematician Srinivasa Ramanvjan (who contributed the superstring theory) attributed his discoveries both to the 'family goddess' and to his mother *who was an astrologer* and had so introduced him to number and to a creative way of thinking about number.[3]

Therefore, reader, please be aware as you read that this book on the esoteric astrology of the West could not exist without the mighty contribution in astronomy and mathematics across millennia from Mesopotamia, Egypt and India.

A further selective process that occurred during the writing of *Astrology: history and purpose* came about as a result of my conviction that astrology is a legitimate practice of the stream of the perennial wisdom

---

2   Algebra was one result of the synthesis.

3   I am deeply indebted to the book 'The Crest of the Peacock, Non-European Roots of Mathematics' by George Gheverghese Joseph (Penguin Books, 1991) for this overview and other wonderful examples.

and that it has practical muscle. This meant that astrology could not be separated from its *mystical* context: it could not be presented as a list of historical eras with cute examples. Thus I brought the historical and the mystical into sync and, as a history, this enabled me to write about the key moments when astrology came under siege and was forced underground along with other communities of the esoteric vacating the turf wars that are the inevitable result of purely historical thinking.

In so confining myself to Western examples where the penetration of a spiritually informed, living astrology is present, some further savage omissions had to be made. One such is the impact of Eastern Shamanism and its ecstatic fruit, which is so much a praise-dance of the Glory; another is the known influence of both Indian and Egyptian esotericism on the Pythagorean School. In a truncated move I have chosen to commence my brief line of enquiry into the Pythagorean school strictly at the point when it moves into the West (Chapter 8).

Finally and critically, I have sought to illuminate the correspondence between the Glorious Earth and the human being – the little Earth – who comes into possession of the Angel Body, as I am certain that the formation of the Angel Body in the human being is the exact correlative to the presence of the Glory in nature, moreover that the Angel Body is the vehicle of the direct sight of the Glory and so the re-enlivening of the world.

Such illumination is the duty of mystical astrology.

Influence and Purpose

The major influences that have met me at certain times of my life that have been integrated *as* influences will be readily apparent here. They can be set out in chronological order as follows.

First, there was my Christian upbringing and schooling, which has resulted in a lifelong familiarity with both testaments of the Bible. Astrology arrived in my life when I was nineteen years old and by my early twenties, when I was at university, I was also studying astrology in my free time. By age twenty-five I began to do charts for friends, gradually moving into professional practice. I joined astrological societies, attended and sometimes presented talks, and conversed with other astrologers, becoming familiar with astrological referencing (and with the astrologer's sense of humour!). Then the New Age movement arrived in Sydney and an intense period of course work and self-examination commenced, a sorting out in various arenas of what was 'mine', what 'projected' and what may be objectively so. The astrology practice deepened as a result.

The teacher arrives when the student is ready. An initial teacher appeared in a person of great knowledge whose presence shocked me into a realization of my own ignorance of the history of spirituality and of human consciousness, and who inspired in me a prolonged period of self-motivated studies of the kind that universities do not provide. As a result, my studies and explorations in the spiritual circles in Sydney brought me into the class of Fourth Way teacher Dr Phillip Groves and into contact with a group of genuine seekers. The impact of Fourth Way (or Work) principles upon prepared ground opened other pathways for reflection, and it was during those years (after an ongoing astrological practice of fourteen years) that I began to wonder what astrology might be *really* and whether it may have an inner purpose compatible with some key ideas of the esoteric circles.

I was at the age when networks expand rapidly, if one is searching. I was enriched by conversations, study weekends, recommendations, teachers and books. Of the many influences that crossed my path at that time, the journal of the Temenos Academy and the writings on the inter-testamental period by the British scholar Margaret Barker stand out. The former made me aware of the visionary knowledge of the World of the Imaginal and of the profound scholars of Islam; the latter linked me firmly to facts about the history and role of the feminine face of the Divine Principle in the Christian heritage.

Particularly, I began to recognise in astrology a contextual, feminine way of prioritizing in the world and, as it became clear in the public domain that Mother Earth is in trouble, I saw clearly that the mental pathway opened up by a practice that emphasises context and belonging may be of extraordinary help. For astrology is a purposeful art. The practice of astrology is like a recurring festival that celebrates each human story at the level of symbol. The understanding and application of the symbols can build the character itself to a threshold moment, when escaping the dangers of contemporary history may indeed become possible.

Finally, it is axiomatic in the study of astrology that there is no such thing as a good horoscope or a bad horoscope in potential. There is only a conscious temple or an unconscious cage. The horoscope is one's temple and the transits are moments of visitation by the wandering Wise Angels.

Make them welcome and ask them questions. You won't be disappointed.

THE GLORY AND HER WHEEL
Chapter One

There is one glory of the sun, and another glory of the moon, and another glory of the stars: for one star differeth from another star in glory.

– 1 Corinthians 15:41

Maybe there was once a word for it. Call it grace.
I have seen it, once or twice, through a human face.

– Judith Wright

The Sap of the Sacrificed God

The most pressing problem of our age is the recovery of our perception of the Earth as Sacred Vehicle. For too long, the Spirit itself has been understood (if referred to at all) in the public Western discourse as transcendent, 'up there', separate from practical matters and invisible. What is required in these times is a renewed immanent and embodied Spirit, one that resides within.

The embodied spirit has been represented in myth and religion since earliest times. Interestingly, in the Norse myths and those of the Hopi Indians, the immanent spirit comes to us largely *because the transcendent god has been sacrificed* either by death or by its own recognition that, to bring the Earth to life again, its own transcendent status must pass; and so the god sacrifices itself voluntarily. The Hopis tell us that after the sacrifice, Spirit resurrects as plants and animals. When the plants are eaten to nourish us, we become part of the plain of Glory. It is a beautiful tale touching upon the understanding of food as well as the necessity of sacrificing the worship of the distant in order to reclaim the cycle of the embodied.

The Sufis of the Islamic tradition tell us that the spiritual preparation of a human being is like the correct preparing of food and that, through such preparation, we gradually become fit 'food for God'. This is not about man or woman being plucked up by a superior transcendence but rather a description of the acquiring of a state of being whose very existence *is* sustenance for, and participation in, living spiritual truth.

Essentially, we ripen.

In some stories there is great resistance to the sacrifice required. The consequences are quite graphic: the Aztecs describe a ferocious earth goddess who weeps with longing to eat human hearts and will not bear fruit unless she is watered by human blood! This appears to be a tale about the causal roots of our self-destruction.

The Book of Proverbs in the Old Testament is a celebration of the worldly mission of the feminine partner of Yahweh, who is known in ancient Jewish theology as the wisdom spirit (ruache) that rushes forth and *fills the universe*. Jesus had something to say about her emanation, the Holy Spirit. Jesus, who taught forgiveness above all things, tells us that the only thing he will never forgive is the denial of the Holy Spirit; in Aramaic (the language of Jesus) Holy Spirit bears the feminine form. In Latin, the Holy Spirit becomes masculine. Sometimes the truth is lost in translation.

The beauty and power of the Earth, its life force, glows as the sap of the sacrificed god, like an overflow halo which surrounds the Earth. We need to summon into our awareness once more this Glory of the World, the nimbus of mysterious light that emanates from within every living thing: eucalypt and wombat, wildflower, worm and microbe of soil, as a continuum of vitality connecting and embracing the whole life cycle.

The loss of our capacity to see and so honour the Glory means, quite simply, that we are starving. The compensations for our collective emptiness are at the root of every problem besetting mankind: the greed, the addictions, the encroaching ugliness, the numbing of our responses, our unease, the chasm between the materially rich and the materially poor everywhere, the frightening statistics regarding the abuse of women and children, the heating up of the climate. Therefore it is vital that the pre-conditions that might enable us to become once more a witness to the Glory be re-established. We need to know ourselves as a part of the glorious web.

We need to be overcome by beauty.

Moments of Glory

Some years ago, when a friend of mine told me that he had a terminal illness, he remarked that a death sentence causes everything to fall away, all of one's history, except for certain events and people that come up as light. The moment of presence, which occurs as one's death sentence is announced, becomes also the moment when one discovers the true story of one's life. For there is a path of gold in the lives of individuals, as well as in the history of civilizations, that marks out the moments when we have been injected with this mysterious light, when there has been a meeting between human potential and the corresponding Glory. Great cultures arise from this if there is a critical mass of men and women seized by inspiration; or at the personal level an enchantment occurs, a mysterious moment of inexplicable grace. These moments arise randomly, when we are happy for no reason in particular, indeed even in the midst of serious negative struggles. Suddenly, there it is – a great peace, a reconnection with some eternal context, a sharpening of the senses, the recognition of beauty.

Once, when I was travelling alone by car in the springtime, I wondered to myself what the world would look like if the Glory used my eyes and my ears, if my own static was put aside as it were? Immediately, the entire countryside before me brightened and seemed to lose its shadows in an overpowering and empowering brilliance, as layer after layer of colour intensified and refined its spiritual tuning.

Let's face it – we have *all* had similar experiences. Therefore, why do we still doubt the immanence of spirit (if we consider this at all)? Why do we not hold tight the real questions that may unlock this beauty; for the Glory certainly responds to the questions that are worthy of her. Why then do we return our hearts to stress and disbelief and our lives to the trivial?

I believe that our failures relate, in part, to an ongoing collective tendency to sleep at the very moment when the strongest wakefulness is needed (Uranus in astrology), as well as to a failure to acquire the will to stand firmly in the other values without flinching (Pluto) and to the invisible nature of the feminine principle itself (Neptune).

Our first duty to the Glory in these times is to *validate internally these experiences* and to struggle most seriously with our human forgetfulness. We must treasure and guard the moments when we have known the glow of this life essence and everything else has just fallen away. For we are all 'terminally ill' and ought not to need such a shock to begin to trace the line of Glory in our own history and to recall the individuals who have carried these moments for us.

This is where a latent power in astrological thought becomes relevant and compelling.

Geo and Helio

In the ancient world, there were several schools that acknowledged the Sun as the centre of our particularly part of the universe and knew that the Earth proceeds around the Sun. Aristarchus of Samos taught heliocentrism in the third century BC. Al-Biruni researched ancient Indian traditions and wrote that certain Indian teachers maintained that the Earth revolves around the Sun.

Nicholas Copernicus himself (1473-1543) cites the Hicet the Pythagorean in his preface to his seminal publication on heliocentrism. He also remarks that the Italian Pythagoreans affirm 'that the centre of the world is occupied by the fire'. Copernicus also owed a considerable debt to the

Arab astronomer Nasir al-Din al-Tusn (1201-74) whose ideas (which fall just short of putting the Sun at the centre of his model) are incorporated into the Copernican model.[4]

Heliocentrism adds to our knowledge of material facts but adds nothing to the spiritual knowledge that happens courtesy of our refection upon the meaning of our own immediate and observable state. In fact, heliocentrism alienates the human being from his or her own perceptions.[5]

In contrast, geocentrism works from our legitimate subjective experience, as human witnesses of the sky, of the flatness of the ground[6], of our looking backwards and forwards in space and what these differing viewpoints might tell us about the symbolism of space and time. Geocentric astronomy is known as traditional astronomy, and of course it is traditional astronomy that forms the basis of astrology itself.

Our Lost Capacity – The necessary Astrology

Last year, when I began to consider writing a book and saw that my concerns about Mother Nature and about respect for astrology might be connected, I began my search by wondering about the manner in which data becomes *alive* in the mind, the thinking pathways that lead to the treasure. How do we come to see the sacred dimension of the earth, beyond theory? It dawned on me that astrological thinking does indeed

---

4   Then the Western historian popped up and announced: 'we discovered that the earth revolves around the Sun. Copernicus uncovered the truth!' All we uncovered was our own vulgarity.

5   Traditional thinker Titus Burckhardt writes: ' it is completely useless to know that [the earth] is round, since this knowledge adds nothing to the symbolism of appearances, but destroys it uselessly and replaces it by another which could never express the same reality, all the while posing the inconvenience of being contrary to the immediate and general human experience. The knowledge of facts do not have, outside the interested scientific applications, any value.' From 'Mystical Astrology According to Ibn Árabi' by Titus Burckhardt, Beshara Publications England, 1977.

6   The author wishes to state that she is not a member of the Flat Earth Society

create a precondition for such revelation. It lays the table, as it were. For the astrologer, the tipping point arises when a funny little light begins to visit the mind during the grand translating exercise that is a session with a client or friend, a session that translates the practical story of the individual through the space of the multiple astro-languages and back to the client in a manner that locates his or her life as purposeful and *connected*.

Around the world, indigenous peoples have always recognised that the land is a series of frozen stories to be replenished by human witnessing rituals.[7] Here, the spirit is utterly immanent: the rocks live, the birds carry messages and the flesh houses the urgency of the Glory to be known. Astrological thinking has an automatic compatibility with the charge that is the earth's nimbus: it is at home there, so to speak, and rides upon its vitality.

There is a particular pleasure that always accompanies a glimpse of real knowledge. Pseudo-satisfactions retreat. Less is needed. An enigmatic happiness increases. Over a long practice there is no doubt that astrology morphs into a total language with which to approach the world, a language unlike any other, being a fusion of storytelling, symbol, poetry, geometry and number, and which must be circular, linear and vertical all at once if the study is to begin to throw up its secrets. Emotionally, it is a joy and deeply satisfying; and its contemplative potential is infinite.

Of course, humankind cannot recover a language it does not know it has lost and in so many public arenas the blind continue to lead the blind. We have all but lost our ability to locate another story for our culture, one based on the sacred and the living; or rather we have by neglect weakened our capacity to *take the effort seriously*. No wonder, as they progress, astrologers pass into a state of wonder tinged by

---

7   In Australia, indigenous people developed their own version of the zodiac, with native
     creatures such as the emu reflected in the star patterns of the sky as part of a holistic
     mental world.

anxiety, at a loss to express the value of the gift.[8]

Astrologers must find again the confidence and respect for astrology as an authentic spiritual doorway. We must introduce the client to the chart as a temple that can, rightly understood, support growth and we must introduce those who are ready to the true purpose of the transits. We must refuse to emphasise 'prediction' which is a corruption of astrology and a consequence of gutting the life and leaving the husk. In fact, I have called our way of thinking the 'Prophetic Mind'.

What does this mean? It is a manner of mining inside data until a precise image or single word appears. If it is a true image, it will be accompanied by a sort of thickening in space that says 'this is not thinking about, this is the thing itself'.[9] Such images (and words, usually multi-levelled ones such as 'penance' appear) can also be put together to create a storyline associated with the horoscope.[10] As the chart is a circle (a spiral really), the point of entry for the story can be anywhere.

As our individual stories-in-time search the stars for meaning, we may draw down that meaning as a *capacity* to witness for the Earth. We must serve the Glory and rightly understand her wheel. For this, there is one preliminary needed.

---

8   The anxiety arises as one is embarrassed to admit to those with social status that one is an astrologer and that this practice is the source of one's perception. I have even had some react with fright and step back when I have announced my profession in a public place.

9   Perhaps a very, very simple example will serve. One may have a client with a conjunction of Mercury and Neptune in the sign Libra. The image of, say, a wind chime is one authentic marker of such a conjunction and perhaps this image may come to mind during a reading. Now, does this mean that the client needs to be asked about wind chimes? Well, yes and no – remember, astrology occupies a liminal space. My own experience is that the mind divides and images and words appear which may or may not be communicated to the client. What is important here is that it is an absolute language with its own integrity. It is not to be taken literally ('do you own a wind chime?'), although it may; nor is it a psychological simile ('your mind is like a wind chime'), though this may also be true.

10 I went through a period when I believed these stories to be 'karma', which approach, like psychological interpretation, seems to be an important stage of development. Over one's lifetime, many paradigms may be used to read a chart before the prophetic keys are given.

Our Lost Capacity – the necessary Angel Body

Any true reading of an astrological chart contains, by implication, a suggestion that the client is an angel in potential. Think about it. Any astrologer worth their salt will make a sincere effort to raise the client's perception of the spiritual patterning and potential in his or her life; and the astrologer will be on the lookout for key statements from the client through which the seed of the soul may be making itself known.

This can never be achieved in just one reading and all astrologers know how much they treasure the people who return each year to examine the year that has gone and to discuss their plans for the following year in the light of astrological timing and possibilities. Yes, we must commence with basics, with difficulties and contradictions in the chart – why the client has come in the first place – but there must also come a point when the client is invited to become serious about their spiritual growth and the proper utilization of astrology. For a few, we may talk about the Angel Body.

The Angel Body (known to some as the Astral Body, to some the Resurrection Body, to some the Kesdjen and beyond) is the result of the Glory emanating from the human being becoming 'set' and a permanent feature of the man or woman, like another skin or the additional arms holding objects of power seen in statues of Indian gods. The creation of the Angel Body is achieved by effort; and *the transits, particularly those of the outer planets, are the triggers for additional work toward this goal.*

Some features of the four stages in the development of the Angel Body will be explained in chapter 5, and some insight into what is being harnessed during transits to assist the Angel Body (should we have such a goal) is covered in chapters 13 to 15. Here it is enough to point out that an inner reversal of one's life emphasis both motivates and accompanies the construction of the Body. This reversal can occur due to stressed circumstances – it can be foisted upon us, so to speak – but in reality it must become sincere and voluntary for the Angel Body to begin to grow without distortion. We need to change our aim and we must know right from the start that this will be hard – there will be unanticipated sacrifices and no guarantees.

The preliminary result of working to build the Angel Body is an occasional increase in the capacity to experience ('see') the Glory. Later we may be authorised to speak on her behalf, to state her requirements without fear because the developing Angel Body is a part of the glorious continuum itself. The work gives it a muscle of consciousness and a new language, and the entire process is supported by astrological cycles, as we shall see.

Culture, History, Exile

Astrology cannot be separated from the context of the history and culture of most of the ancient world: it was an assumed element in all religious thinking. There are some lovely books on the history of astrology[11] – and astrology in history – and I do not need to go over here the same factual material. What is presented below is a sampling of some of the events that erupted when prophetic thinking penetrated a culture via individuals with the prophetic capacity – or when the culture itself possessed this - when it was apparent that the Glory and her wheel were somehow afoot; for it is the infusion of glory that *makes* a civilization and a lasting culture, that tells its history as a touchstone for the memory of the race. It is with this aim in mind that the traces of wisdom carried by the priests of the Western European megalithic sites and by the Sabians of the Middle East, shrouded in mist, are told, how the glory trail flowed from the Pythagoreans into Italy and from the First Temple Jews into the Book of Enoch (lost for 1500 years) and the AD Neo-Platonist fusion that became known as the Hermetica. We will resurrect

---

11   Two contrasting books are 'Astrology for the Age of Aquarius' by Jan Kurrels, Parkgate Books 1997, which has a clear, basic introduction to our history and simple, clear astronomical diagrams; and 'The Fated Sky' by Benson Bobrick, Simon and Schuster 2005, which is full of interesting anecdotes and historical detail.

Christianity in the light of the Angel Body and bring these things into the Renaissance, when the astrologer-magicians of the times impacted on art and science. In all this, astrology will be our companion.

We will pass at last into an analysis of the means and grand mistakes that caused astrology to fall and materialism to rise, how our capacity for enchantment diminished in direct proportion to her loss.

For, one could say that astrology is in exile – and she is in good company. She sits with seminal figures from history, those who have either purposefully sought for or been graced mysteriously by the Glory. This sense of being in exile is the companion piece to the practice of astrology. Of course, the Glory is never in exile: it is only the normalizing of the loss of her that makes it seem so. It is our expectation of a renewed visitation by her as a state of grace that encourages us to practise the contemplation of her wheel.

OF SABIANS AND STONE ASTRONOMERS
Chapter Two

*(The) implicit primordiality of astrological symbolism flares up in contact
with spirituality, direct and universal, of a living esotericism, just like the
scintillation of precious stones flares up when it is exposed to light.*

– Titus Burckhardt

Anatolia

Astrology has its roots in a way thinking perfected in archaic times,
back even as far as the late Neolithic Period. By the middle of the
third millennium BC, this way of thinking had been inherited by the
Sabians, a loose term - about which there is still some debate - denoting
a number of 'star-worshipping' cults spread across (modern day) Iraq,
Syria and Turkey. The focus of Sabianism was in Harran, which is now
a ruin in southern Anatolia, Turkey. The Harranians had inherited their
tradition from the ancient Syro-Babylonian cults, as part of the great
underground stream of esoteric knowledge stretching back into the
mists of Shamanism. These in turn had inherited religious forms drawn
from the ancient gnosis.[12]

---

12 Gnosis draws on a supposed capacity to perceive the mystery within spiritual matters
directly, in contrast to intellectual considerations, theology, logic etc. It is not to be
confused with Gnosticism, which was a formal movement within Christianity.

The inscription above the Great Temple at Harran read: He who knows himself is deified.[13]

What does this mean? It seems to imply that mere self-knowledge will turn one into a god of sorts. What does 'deified' mean in this context?

The relationship between self-knowledge and progressive deification is one of the great mysteries that astrology seeks to unlock. It is our inheritance and the Sabians are our ancestors. Their revelation is our revelation *in potential*.

To understand why our astrologer-ancestors first surface in Southern Anatolia, we need to track back to several of the great and mysterious megalithic sites of Western Europe. These were suddenly and inexplicably abandoned by their inhabitants beginning at about 3000 BC, after which time a perceptible trail of emigrants can be found perceptible trail of emigrants can be found, snaking particularly from the British Isles to various bases across the Middle East. Southern Anatolia was one place that received the emigrants.[14]

The new arrivals to Anatolia were led by priest-sages (Magi). These carried with them knowledge of astrology, astronomy, medicine and control of the weather, and they summoned the spirit world. In short, they carried shamanic skills. They wore high hats: archaeology has uncovered a female in such a high hat in the region. A vast exchange of knowledge took place as the infusion of an advanced megalithic approach to temple astronomy and reincarnation flowed into the entire region.[15] In fact, it is only in Western Europe *and* in the Middle East that circles of standing stones are found.

---

13   This was purportedly a saying of Plato and may have been placed above the Great Temple after Greek thought and later Neo-Platonism reached Harran. Another translation reads simply: Know Thy Self.

14   The Sumerians settled between the Tigris and the Euphrates, bringing their own calendar and their astronomy. Here they instigated mathematical activities across the region. Other tribes inhabited what is now Israel, began to call the place Ca-na-na-um and as Canaanites, to worship Venus as Ashtaroth, the mother of the God of Dawn and the God of Dusk.

15   A similar understanding of priestly reincarnation entered Greece from Ancient Mongolia at a much later period, around 600 BC.

Therefore, if the birth of Western astrology arose in Anatolia as part of a river of knowledge that included megalithic astronomy, what attitudes and philosophy – what *mind* – did these immigrants bring with them?

Stone

The megalithic stone circles of Western Europe have been variously carbon dated back as far as 5500 years before our present era. Some sites in Scotland and Brú na Bóinne in Ireland are one thousand years older than Stonehenge and were built before the use of metals began. This was a time when authority was held by the astronomer-priests.

The standing stone temples were geometric shapes and were precisely planned using three-dimensional polygons in stone and templates based upon the division of a circle into seven or nine equal parts. In some sites, henges were pressed into the surrounding ground and these provided an adjusted viewing platform or a water-filled star observatory using the water to reflect the sky. Archaeological research has revealed several kinds of geometric shapes – all roughly circular, hence 'stone circles' – in the plan of the temple observatories. These are the circle, the ellipse, two kinds of 'flattened' circles (as if the top has been pushed down slightly)[16], egg shapes and an egg shape with a semi-elliptical end.[17] The mathematics involved in the laying out indicate that these people living thousands of years BC would have had to have grasped 'Pythagorean' triangles, square roots, right angles and fractional adjustments long before algebra was ever conceived!

---

16   This appears to have enabled the mathematics of the site to avoid irrational numbers in its construction, an adjustment which enabled the uncertainties of the 'irrational' mystical realm to penetrate into the 'earthly' stone observatories.

17   For a clear explanation of the geometry of the shapes and how they would have been measured out, see Keith Critchlow, 'Time Stands Still' New Light on Megalithic Science, Floris Books 1979, 2007. The shapes themselves were uncovered by Professor Alexander Thom; see 'Megalithic Sites in Britain', Oxford, Clarendon Press, 1967.

The stones from the sites appear to have been selected *individually* based upon Shamanic perception of their suitability as power objects plus a profound vision of the beauty that would result when they were set together in the desired astronomical arrangement. The stones *selected* were charged[18] to enable them to remain for a time 'alive' and the repository of divinity.

How were they charged? As there was no metallurgy, what would they have used: invocation, timely weather patterns? Perhaps the astronomer-priests possessed *in themselves* some magnetism powerful enough to charge an atmosphere if they so willed it? However it went, through this charge and by subsequent invocation of the Beautiful, they were arranged as both a temple and an observatory. In their function as temple, the stone circle affirmed the continuum between man, earth, cosmos and sacred breath as a permanent axis; and as observatory orientated to the planets, the sites allowed the precise calendar corrections so necessary for agriculture and for the setting of the rituals that made agricultural practice a simultaneous spiritual activity.

The number of people working to achieve these sites had to be phenomenal, as this was a time when people were old or dead by the time they reached twenty-five years of age. In addition to the astronomical and mathematical knowledge, there had to be a collective motivation so certain, so respectful and inspired, that the modern mind would be at a loss to comprehend it.[19] Their motivation seems to have sprung from a manner of thought that was prophetic-sacred and scientific at once – hence, astronomer-priests. This (probably Shamanic) approach

---

18   Interestingly, there is a book by the American Work teacher E.J. Gold, in which he claims that the charges in some ancient stone objects can still be accessed. Gold says that much archaic knowledge was stored in the stones as a legacy for those able to sense them. See E. J. Gold 'Visons in the Stone' Journey to the Source of Hidden Knowledge, Gateways/IDHHB Inc, Nevada 1989.

19   Keith Critchlow points out that our modern era is the first to be 'unwittingly attempting to conduct society without the metaphysical or spiritual dimension' known to be a fundamental component of every traditional culture in history. See Critchlow op. cit. p.12.

to knowledge invited living symbols to break into Mind as light; after they 'returned to Earth', they were able to locate the symbols in nature and charge them with power. Thus the megalithic monuments were constructed from symbols and, as temple, orientated to the movements of the wanderer planets, particularly to the cycles of the Sun, Moon – and Venus the Morning Star.

This was the Stone Age.

Metal

Many finely chiselled stone objects have been found by archaeologists working in Western megalithic sites and around them, where underground rooms for workers' residences and food supplies were located. At sites in Scotland, such as Skara Brae on Orkney off the northern coast, small 'spherical-ised' stone polygons carved from granite and other hard materials have been found with precise measurements such as seven equidistant protrusions from the smooth stone.[20] These stones would have had to be made without the help of metal tools as we depend on them – no chisels, hammers and the like.

Much later, after metals came into common use across the world, Skara Brae was abandoned. The departure of the inhabitants of Skara Brae has been dated at about 2655 BC.

Now, well within one or two generations after the exit of the inheritors of stone from Scotland, Giza was established in Egypt and work on the Great Pyramid was commenced; and it is on record that there was a tool used to shape the stones of the Great Pyramid that was *not made of metal*, despite the use of metals in that place and at that time. Furthermore, sixteen hundred years later when King Solomon built his temple in Jerusalem, the use of metals was *banned* in the construction of this

20 This indicates that the carvers had a clear conceptual grip on polygons a long before the time of Plato, whom we credit but who inherited this body of knowledge during the time when the ancient teachings were being set down in writing.

powerful spiritual site. Why? Whatever the answer, we have a theme of no metals – initially through the historical setting but later through deliberate edict.

Now, there is a correspondence between the planets 'in heaven' and the metals 'on earth'.[21] Such correspondences are used in alchemy and known in Hermeticism as part of the law of 'As Above, So Below'. Therefore, if the use of metals was banned in the establishment of such critical places, erected with geometric purity as pilgrim site, shrine and transmitter of God,[22] it may be that it was planetary influences which were being avoided because something was known about *the way these influences may operate to distort the purity of the project*. After all, there is a powerful spiritual teaching that arose in the Middle East after 3100 BC which states that both metallurgy and astrology are 'fallen' arts and that they were brought to Earth by Fallen Angels...[23]

The Chamber of Venus

Anatolia is a place of copper: there is a large quantity of the metal in the ground across the region. Of course, the modern mind will see any metal-loading as a mining and manufacturing opportunity only, a resource to be exploited and extracted in its entirety. However, copper is *also the metal of Venus* and those who arrived in Anatolia just before 3000 BC brought their new-era knowledge of metallurgy and their astronomy with them, while other refugees from the megalithic sites spread the worship of Venus (in her earthly forms) across the region. Other far-flung emigrant groups kept bases in Anatolia. Why? Perhaps

---

21   Sun-gold, Moon-silver, Mercury-quicksilver, Venus-copper, Mars-iron, Jupiter-tin, Saturn-lead.

22   Stonehenge, The Great Pyramid and Solomon's Temple are said to be the three primary man-made witnesses to the Glory of God, which claim is supported by the numbers used in their proportions.

23   This is the Enoch teaching, finally written down between 200 and 100 BC as The Book of Enoch. See chapter 3.

we drift where the soul in its cosmic context senses a compatibility, and that which we dance upon radiates a call-and-response with our own cosmic tribe: Venus-copper-Venus-copper…[24]

Venus was vital to the Anatolian mysteries. Why?

The astronomer-priests of Western Europe, those who had carefully observed the rising and setting of the planets and the stars of first magnitude, knew that the orbit of Venus transcribes a pentacle in the sky,[25] the five 'angle points' being separated by eight years each, giving an exact Venus return after forty years. This is when the light of Venus will angle precisely as it did forty years ago and so it can be anticipated and captured, for instance, by a stone passageway leading into a chamber. This Venus-orientation is apparent in several important sites in Britain and, in true Stone Age style, the Venus-light served both a practical and a spiritual purpose.

Astronomically speaking, Venus set the calendar. There are three essential calendars necessary to enable a society to function. The proximate lunar calendar allows us to predict the tides; the solar calendar gives the length of the days and the setting of ceremonies; and the sidereal calendar – exact time according to the stars – is used for agriculture. Over time, these calendars drift apart creating three conflicting 'clocks'. However, every eight years they come together within minutes of accuracy and every forty years they are accurate to a matter of seconds. It is easy to see how the tracking of Venus-light would hold together the strands of society.

Venus in her role as harmoniser became the reconciler of time itself.

The pentagon is one of the fundamental invisible floor-plan designs

24 Certainly indigenous people have known their country well enough to set sacred sites at points of great elemental power, and it is both sad and amusing to observe incoming capitalists claim resources (such as uranium) from these sites and accuse the aboriginal people of creating a convenient myth to stop 'development'.

25 From the point of view of the earth, Venus loops toward us five times in a cycle. This is also illustrated commonly as a five-pointed star, and as a five-fold flower or an intricate and beautiful skein of thread.

identified in the construction of the stone circles.[26] At Bryn Celli Thee on Anglesey, a notched pillar may be found inside the chamber itself; the topmost notch is touched by Venus-light only during her return, the other notches referring to the solar equinoxes and solstices. Likewise, at Brú na Bóinne in County Meath, Ireland, solstice light and Venus light penetrate right into the chamber via a slit in the lintel and fall on the rising floor of the inner passage. This light must have had another purpose because light penetration into a chamber is not necessary to identify a Venus return and so set the calendar.

The five-pointed star is an ancient symbol. For many ancient communities, the five-pointed star was a symbol of regeneration. For the ancient Jews it represented rebirth, and for the Egyptians too Venus stood for – *was* – rebirth and this knowledge they had in common (once more across a strange connecting tissue) with the folk of the stone circles. All knew the science of Venus as the science of rebirth. It is not hard to see why they did. For part of the year, Venus is the Morning Star and rises before the Sun; yet during the other part of the year, Venus is seen to rise in the evening as Evening Star. The ancients would have observed this rising-setting-rising-setting, all within a five-fold star pattern, the star that is the abstract of the human body with arms outstretched; and so Venus became the rising and setting of the human being, the eternal return of soul to embodiment, set within a great cosmic beauty.

It is always tempting to interpret 'rebirth' as the reincarnation of individuals and to leave it at that; and perhaps Venus was seen as a midwife and blessing to such a process.[27] However, Venus as rebirth has a finer meaning than this. It was during the period of the great revival of Hermeticism and its impact on Christianity in the 15th century that

---

26 Keith Critchlow has done a thorough analysis of the pentagonal shapes identified by Alex Thom, particularly of Moel Ty Uchaf in Northern Wales, and has suggested the latter was dedicated to Venus.

27 Recent research has indicated that births associated with the ancient priesthood may have been timed deliberately to take place during the days of the Venus return and that this was one function of the megalithic chambers.

bishop-astrologer Marcilio Ficino that *Venus signified the Angelic Mind itself.* He taught that the Angelic Mind is that 'which by innate love is compelled to contemplate the *beauty* of God' (emphasis added) and through its rulership of the generating power of the World Soul (the Glory) recreates this beauty in the forms of nature. The Angel Mind invites us both to care for the Earth *and* to remain in harmony with the subtle energies of the universe. The joint rulership of Venus over Libra (contemplation of harmony) and Taurus (earthly manifestation of beauty) helps our understanding of the inevitability of this connection.

Love itself is forever reborn. The world as beauty is forever reborn.

Therefore, the megalithic masters arrived in the Middle East and in Anatolia, and with them came both science and a philosophical, integrated beauty. Whatever their motivation for abandoning their former homes and whatever the intention of the Glory and her Wheel of Correspondences, we find ourselves at last at the source spring of Sabianism, which movement possessed their own geometric temples, charged stones and mystery.

The Temple of Rejoicing

By about 2300 BC, our particular awareness of the Sabians of Harran as inheritors of a wisdom stream in Anatolia seeps into history – and it ends abruptly when their temple is destroyed after AD 1032-33.[28]

The Great Temple of the Sabians in Harran was known as the Temple of Rejoicing and was dedicated to the Moon God Sin. It was the time of the Great Age of Taurus and, in the astrological hierarchy, the moon is exalted in the sign Taurus and written as a glyph that is both the new moon and the horns of a bull. In Mesopotamia, the lunar horns of the bull were a symbol of male fertility, while the full moon indicated

---

28   There is some dispute about the dates and causes. For example, it has been said that the Temple was destroyed by the Egyptian Fatimid dynasty in 1032, while others claim that the governor of Harran himself destroyed it in 1081.

the moon's feminine aspect. Bull sacrifices to the Moon God Sin were common.

Taurus rose at the vernal equinox from approximately 4000 to 1700 BC.[29]

The Moon God Sin was thought to reveal the secrets of the Cosmos to those who sought him out and to give oracles by serving as *a* reflecting *tablet* upon which the divine writ could be inscribed. Nergal, the twin brother of Sin, was the ruler of the dead and linked to Saturn.

So, since the 'Sabians'[30] are astrology's most recent historically verifiable ancestors in Anatolia, what do we know about them?

The Sabians were recognised by those around them as a well-educated group with an authentic tradition. (After the Muslim conquest of Turkey, an Islamic university was established in Harran.) However, much as they were acknowledged as a group, it seems that their beliefs were always hard to pin down. Writers and scholars have positioned them 'between' other traditions and say that there was 'heated debate' among Orientalists as to their true nature. They seemed to resemble Jews and Christians and Magians and yet be none of these; and the imams were unsure of them and could not obtain knowledge of them. They appeared not to exactly reject the prophets of Islam – but neither did they regard it as obligatory to follow them. Astrologers will recognize this attitude all too well!

They have also been described as *a sect of Angels.*

What else do we know?

We know that they declared one indivisible unity (or god) as the

---

29   By about the sixth century BC the equinoctial sun, now in Aries, appeared precisely at Aries' brightest star, Hamal, which would have assisted the Babylonian astrologers in their quest to finalize the order of the zodiac as we now know it with Aries the 'first' sign.

30   We shall refer to the astrolators of Harran as Sabians, even though there remains a dispute in scholarship over the precise use of the term. Apart from the Sabians of the Qur'an and the cult in Harran, there were so-called Sabians living in India, Central Asia and Syria: these may well have been descendent clans of the original stone circle emigrants.

supreme principle, even if they did not express this as a religion. We know that they existed before Abraham and that Abraham spent significant time in Harran, seeking to convert the 'sabi'ah' there. We know that the Sabians took the biblical Noah-of-the-flood[31] as one of their key prophets. Now, there had been a literal flood in the time of Noah; however, sacred books are written as the story of the Glory moving *within* events, not purely as a record. The flood is a story also of the effort by the conscious to rise above the flood of ignorance that is always present to swamp and distract: when we rise, we preserve the truth. We can surmise then that, since Noah built the ark to preserve esoteric knowledge from the rising waters of ignorance, it is probable that the Sabians sought gnosis and practised discretion.

Certainly submersion in water was one of their rituals (although this was a common practice). One theory has it that the root of the word 'sabian' is Syrian and means 'baptiser'.[32]

We know that they studied the Psalms of the Old Testament, one of the OT's 'revealed' books.

When I consider thy heavens, the work of thy fingers,
the moon and the stars, which thou hast ordained;
What is man, that thou art mindful of him? and the son of man,
that thou visitest him?
For thou hast made him a little lower than the angels,
and hast crowned him with glory and honour.

– Psalm 8

---

31   Prophet Noah is a feature of The Book of Enoch, recovered in Ethiopia in the nineteenth century and later at Qumran, about which there will be more discussion.

32   Others say that the real Syrian root is sba, a verb signifying 'desire', 'to desire the knowledge of God'. Still others claim that Sabianism is from Sabi, a son of Methuselah. In religious history, Methuselah is the son of Noah. The term 'saba'a' can also mean 'he changed his religion'.

Later, after the transition from the Age of Taurus to the Age of Aries, the Sabians built temples dedicated to the Angels of the Sun, Moon and the five known planets of their day, and they had developed a system of correspondences to these planets: colours, gemstones, days of the week. They performed their Temple enactments to the appropriate planetary angel each day, in robes of the appropriate planetary colour, and they used a lunar calendar to time their cyclical activities and festivals, no doubt inherited from the Moon God. Most of their rituals have been lost, although there have been attempts to reconstruct them.

It is terrifically hard to substantiate any description of the form and timing of rituals.[33] However, there is a strong case for the relative authenticity of texts known as the 'Epistles' of the Brethren of Purity, which describe some features of Sabian ritual. The Brethren were a 10th century AD offshoot of Shi'ite Islam that sought knowledge through 'divine wisdom granted through revelation' and scholars have been able to show that rituals and teachings credited to the Sabians formed the basis of the perception (or gnosis) of the Brethren; and the Brethren themselves claimed that the Epistles had been written before the ninth century period of translation in Mesopotamia[34], making it more likely that they really were an undiluted record of the Sabianism of that time.

Then, what might a Sabian ritual have looked like according to the Brethren?

---

33   The problem is that the Sabians have been constantly reconstructed in the language of any incoming spiritual movement. They have been reinterpreted by Babylonian, Assyrian, Jewish, Greek, Roman and Syriac Christian generations and by Hermeticist, Neo-Platonist and various Muslim ideologies, many of whose adherents viewed the Harranians through the colours of their own needs – and found what they were looking for!

34   By the 9th century, the works of the Neo-Platonists were being translated in Harran. Interesting, one of the most important translators of that time, Al-Kindi, also believed that the stars were somehow alive and had souls and that their rays interacted with the earth as spiritual promptings. The ninth century translation period is the time when the Muslims began challenging the legitimacy of the Sabians. Several revelations exchanged knowledge during this period.

The Epistles describe a temple in which Sabian ceremony took place. The temple had been built at an astrologically auspicious time. On the northern wall there were the names of the signs of the zodiac and the symbols of the planets, each made from the metal ruled by the planet (e.g. copper for Venus, silver for the Moon). In front of these were seven white discs, dedicated to the seven known planets. The nearest to the northern side was the moon and they proceeded out from there, ending with Saturn. From the Moon to Saturn, each planet had an increasing number of 'rings'.

Each of the planetary discs held a censer filled with the incense dedicated to each planet. Present at the ceremony were also three boughs of tamarisk, an iron knife and an engraved iron ring and water. Pine cones were also burned as incense.

This setting would be highly suggestive to astrologers. What may be less appealing is the presence in the ritual of sacrifices to demons. Mesopotamia had inherited a Zoroastrian good-versus-evil paradigm and the knowledge of the Fallen Angels. The Sabians believed that evil had to be neutralised by sacrifice[35] as well as good invoked and so the Priest of the Temple wore a ring engraved with a figure with a cock's head (the cock crow heralding the dawn and thus the rising of light), a human body and underworld serpents for feet.[36] It was only after the sacrifices that young initiates were invited to hear the mysteries: one set of mysteries for men and another for women.

There is also a clue about Sabian ritual within the writing itself of source material on Sabian ritual. The word 'pneuma' (plural pneumata, in Arabic ruhanija with endings) is repeated again and again and pneuma is an ancient Greek work for 'breath, air, ka, spirit'. Calming

---

35 The Sabians claimed that one difference between angels and demons is that the latter do not respond to prayer. It may be that the terms 'invocation' and 'prayer' were interchangeable in their understanding.

36 It is at this point that a Plutonian tangle of occult possibility rears its head. This has always attracted a particular kind of mind. It is the author's opinion that invocation and everything surrounding it needs to be treated with great caution.

and working with the breath is common to many spiritual practices and is said to be central to making a space in the mind for light. Many streams of the perennial wisdom were interlocking in that part of the world (around Harran) and the Sabians were ancient enough to have had dialogue with multiple travelling streams of knowledge across the ages.

It may be drawing a long bow to suggest that even some pre-Pythagorean breath practices were linked to the Sabians and yet this may be so. The Pythagoreans lived on Samos (off Anatolia) and had inherited a practice from the ancient magicians of controlling the breath and *hissing and piping in such a manner as to reproduce the sounds made by the stars and by the revolving planets!* The sound was called the Syrinx. Where did this practice come from? Who were the magician-ancestors? Was such a practice an incoming influence on Harran or did the Sabian star-breath impact Pythagoras? Perhaps there were identical practices arising over time in more than one tradition – certainly it would not be the first time such a thing has happened. However it goes, when we link this commonly shared part of the world with the astral foundations of Sabian knowledge and with the repeated phrase breath-spirit in later Sabian sources, it seems (even by intuition) that this hissing was their practice.

The Sabians did indeed use the breath.

One further strange ritual of the Sabian priesthood involved preparing carved heads for divinatory purposes – surely this must have been a diminution and corruption of megalithic ritual in which stones themselves received a spiritual charge from the astronomer-priests.[37]

Finally, we have in Anatolia an inheritance of strict geometric structures aligned to planetary symbolism and motion. Al-Mas'udi (d. 956

---

37 A further echo of the living stone statues theme is found in 'The Winter's Tale' by William Shakespeare, one of his final plays and written at a time of intense revival of magical thought in Europe. This play became in turn the foundation of George Bernard Shaw's 'Pygmalion', where it became a tract on class relations, the member of the lower classes being 'brought to life' by aristocratic education. Finally, 'Pygmalion' was conceived as the musical 'My Fair Lady' in the 1960s, thus completing the progressive trivialisation of ancient knowledge (and yes, 'My Fair Lady' is a great musical…).

AD), who visited Harran in 943, has left a description of the temples and the various groups of Sabians. About the temples he says:

'The temple of the soul is round; of Saturn, hexagonal; of Jupiter, triangular; of Mars, long (rectangular); the sun square; that of Venus, a triangle in a quadrangle; that of Mercury, a triangle inside an elongated quadrangle, and that of the moon, octagonal. The Sabians have in them symbols and mysteries which they keep hidden.'[38]

Therefore, we have a clear line of megalithic tendencies, which have been impacted by other movements and loss of memory over time; we have common spiritual markers such as incense and baptism. We have a practice on the breath that may have been a part of rituals to planetary angels. We have sacrifices to demons, perhaps to prevent them distorting the perception of the participants in the rituals. Perhaps. Scholar Tamara Green indicates the core of the matter when she says that they had as their most important paradigm *the visible heavens as reflecting the numinous will.*[39]

The Sabians had a prevailing sense of the numinous.

Uncertainties and Disguises

Time passes, traditions morph and splits occur. By the 9th century AD, the Sabians may have departed from their roots, particularly under the impact of the 'war' with Christianity,[40] the Greek modifications to

---

38 Mas'udi, Muruj al-Dhahad/Les Priaries d'Or, quoted in Tamara Green 'City of the Moon God' E.J. Brill p. 115/116.

39 Green p. 9

40 The Christians couldn't stand the Harranians and their Sabian way of thinking, and the Sabians reacted to Christianity with a complete lack of interest in conversion. The result was that in the 5th century Christian Bishop moved his headquarters from Harran to Edessa. Much later (10th century), a geographer described the established site of the

astrology under Neo-Platonism and the Hermeticism of the first three centuries AD, but we can't really know, and any incoming religion or philosophical school is bound to write critically about what has gone before: they have an agenda. However, what we can be certain of is the ongoing status of the Sabians as uncertain: the ruling elites did not know how to define them.

There is an interesting and oft-quoted story about the Sabians. The story relates that, in 875 AD, the astrologers of Harran were ordered to convert to Islam under pain of death. Under this pressure some of the astrologers – who had never cared what current form of the One Great Truth they embraced – took the obvious and practical step. Many in this group later moved to Babylon as Muslims while retaining openly their 'Sabianism'. There, they were widely respected at the court of the Caliph.[41]

However, there were others who, confronted with the convert-or-die conundrum, took a more subtle approach. These first consulted a lawyer (!) and established that there was a community called 'Sabians' in the Qur'an and that they were to be tolerated – yet nobody knew exactly what this term Sabian meant. Therefore, these astrologers formally took the name Sabian for their group. While they were at it, they declared themselves also followers of the prophet Hermes Trismegistus and users of the sacred text, the Corpus Hermeticus.

Now, at this point, dear astrologers, this author needs you to grasp the subtle nature of what was a wonderful joke, courtesy of Sabian *attitudes*. You see, the Qur'an Prophet Hermes is none other than the Prophet Enoch;[42] and it was the Book of Enoch *which is claimed to*

Sabian temples as 'a lofty heap'. Given this, this author has some sympathy for the Bishop of Edessa: how would you feel if you arrived to put a neat church with its cross on the top of a hill and found there pagan pieces of geometry – with talking heads – 'on a lofty heap'?

41   A Sabian community and school were founded in Baghdad by the Harranian Thabit ibn Qurra (d. 901 AD).

42   Prophet Enoch or Idris is the Islamic name of the Mercurial revealer-messenger serv-

have preserved the secret astronomical/astrological lore of the mega-lithic inheritance and was a repository of visionary thought in the face of successive waves of dead fundamentalism and its wars. In fact, the Book of Enoch was so disputed by formal religion that it had 'gone missing' at about 200 AD.

Therefore, in claiming Enoch/Hermes as their prophet, the astrologers managed to send up the fundamentalist mindset[43] that was seeking their execution. They claimed the prophet of the living Glory herself.[44]

This supposed story took place at a time when Islam was about 200 years old and the writings of the great Greek philosophers – Aristotle, Plotinus – were being translated in Baghdad, perhaps onto the new paper that had been constructed from rags and that now enabled a rapid dissemination of thought. Critics of the story claim that, as Islam had been living side by side with Sabianism for all this time in mutual toleration, it would be unlikely that they would suddenly issue an edict threatening the community. However, generations change and individuals become threatened by ideas that rival or confuse them. It may have been that the conquering Muslims had come under question about their religion from the new scholars and translators; certainly from that time the inner aspect or 'heart' of Islam began to be developed. When those frightened of change are questioned by the educated, some inevitably look for heretics to burn. It is quite possible that *some* Muslims did in fact pronounce a doom on local Sabian uncertainties.

Whatever, the Sabians relaxed back into the perennial stream of spirit, which underlies all religions and where they were comfortable – and hard to pin down.

The Enoch solution was the final recorded 'joke' by the Sabians. By

---

ing the Divine Wisdom. He is also called Thoth-Hermes (Egypt), Hoshang (Persia) and Apollo and Orpheus (Greece).

43   An educated guess would have them assuming that the fundamentalists would be incapable of seeing what they were doing.

44   The Sabians even made pilgrimages to two of the pyramids and claimed that one was the burial site of the son of Hermes/Enoch. Really, can we take this seriously?

the time the Great Temple at Harran was destroyed in the 11th century, astrology was re-entering the religious mainstream in the West. In the next chapter, we will return to the Babylon of the 6th century BC and map the route by which astrology came to the Western world.

TRACES OF WISDOM
Chapter Three

'Come down, and sit in the dust, O virgin daughter of Babylon,
sit on the ground: there is no throne, O daughter of the Chaldeans...

'Thy wisdom and thy knowledge hath perverted thee;
and thou hast said in thine heart, I am, and none else beside me.

'Thou art wearied in the multitude of thy counsels.
Let now the astrologers, the stargazers, the monthly prognosticators,
stand up, and save thee from these things that shall come upon thee.'

– Isaiah 47:1,10,13

Traces in Sacred History

We have seen that astro-religious thought was present in the Neolithic world and that the temple-observatories both expressed and contained that thinking. We have seen how this knowledge was passed over into the Middle East and particularly into Anatolia; and we have observed a reluctance to convert to any 'religion' on the part of the Sabian inheritors of that knowledge. Now it is time to go back again, this time to the early 6th century BC, to the so-called 'cradle of civilization', which extends from the Euphrates, and to a time known as the Babylonian Captivity of the Jews.

There is great relevance in astrologers knowing about the Captivity, as this time was pivotal in the transformation of many traditional ideas that later impacted upon the West, including our approach to astrology.

These were the generations of transformation, of *a change in emphasis*, and which anathematised an entire way of thinking.

The 6th century BC was a remarkable time for spiritual revelation. Lao-tse and Zoroaster were alive at the turn of the century (the latter crystallizing the notions of monotheism and of the 'good and evil' polarity)[45] and soon afterwards Confucius and Siddhartha, founder of Buddhism, were born. In India, the first Jain Mahavira Jina began his mission. In Delphi we have the oracle, and the great theatre being constructed.

The pivotal Western master Pythagoras, too, was born in the same year that Nebuchadnezzar II invaded Jerusalem and brought about the Babylonian Captivity of the Jews. This means that Pythagoras grew up and received his training across the generations in which the Jews – particularly the Jewish intellectuals who were targeted by the invaders – were held in Babylon. It may mean also that the birth of Pythagoras was somehow necessary just at this time; for the 6th century BC contained such a seeding of profound and lasting traditional thought that one has to wonder if there was some disturbance in the realm of Spirit that necessitated an increase in revelation on earth or if there was an intention. It is movements in the realm of Spirit that is here referred to as Sacred History.

Sacred History is the study of *the lines of living energy which constitute the movement of the Holy Spirit across the enabling Glory of the Earth.*[46] It is nothing at all to do with the ordinary history that tells stories of shifting territories, wars, cultures and the like. Regardless of what *appears* to be happening in our world, the living line of Glory is always there – always – and if one senses that the life has left a tradition or body of knowledge, it is a helpful and wise activity to search for where the spirit may be active.

With this in mind, let us look at this important period.

---

45   These ideas corrupted later revelations, much to our cost.

46   In this writing, such a Spirit is also referred to as The Glory in her moving aspect. Such Glory is an emanation, a state and an activity at once.

Visionary Mind

In the late seventh century BC, Jerusalem[47] was under the spiritual rule of the First Temple. The First Temple priests and laity were driven by what is known as a Vision theology, with a temple 'presence' known as the Shekinah, a feminine figure and an aspect of the Glory; there was the King, in this tradition God's mandate on earth. There was the Prophet Ezekiel and the Second Isaiah (the Old Testament Book of Isaiah is written by three 'Isaiahs'). Yet Jerusalem was also an amazing mixture of tensions and potential internal conflicts. Particularly, there were the law-yers, those disgruntled democratizing bureaucrats who later became known as the Deuteronomists.

Then, in 598-97 BC, Jerusalem was besieged in war by Nebuchadnezzar II and fell. Jews in large numbers were removed to Babylon as slaves. The first deportation occurred on the 16th March, 597 BC, others in 587 and 582 BC. Among the deportees were Ezekiel, the Second Isaiah, many traditionalists of the First Temple and the Deuteronomists: Nebuchadnezzar had deliberately taken the key intellectuals captive. Meanwhile in Jerusalem, the First Temple was destroyed.

One can imagine what these 6th century BC generations in captivity were like. The early exiles would have been in shock and grief, and their religious minds would have been searching for the reason that God had visited this punishment upon them. They certainly would have viewed the conquest of Jerusalem and the captivity as a punishment but punishment for what?

One of the prophets in exile was Ezekiel, a prophet of the First Temple thinking that permitted visions (although his Old Testament book, too, is full of a vengeful God). Ezekiel I describes a very strange vision of angels which he receives by the River Chebar during the Captivity.

---

47   Jerusalem, it will be recalled, was part of the site of the emergence of renewed Venus-worship, following a period of stone circle abandonment. Of course, worship of the Great Mother goes back much further, but we are concerned here with the line of transmission that became astrology as we know it.

'And I looked, and, behold, a whirlwind came out of the north,
a great cloud, and a fire infolding itself, and a brightness was about it,
and out of the midst thereof, as the colour of amber, out of the midst
of the fire.
'Also of the midst thereof came the likeness of four living creatures.
And this was their appearance; they had the likeness of a man.
'And every one had four faces, and every one had four wings.
'As for the likeness of their faces, they four had the face of a man,
and the face of a lion, on the right side:
and they four had the face of an ox on the left side;
they four also had the face of an eagle.'

– Ezekiel 1: 4,5,6,10

The figures ride upon wheels within wheels and 'the spirit of the
living creature was in the wheels'.

Astrologers will recognise immediately that Ezekiel received a vision
that, of all the astro-mythological examples in the skies of the northern
hemisphere, included only the imagery of the four fixed signs of the
zodiac. Ezekiel's vision may as well be a description of an archangel
ruling the astrological quality of fixedness – a sort of Mode Angel.[48]
Ezekiel had a clear vision of figures that fused the four fixed signs as
angels. Why?

Now, it was in Babylon and at this time that astrologers were working
toward the putting together of the final form of the twelve sign zodiac
system, against a background of thousands of years of astro-mythological development, courtesy of the stone circle immigrants. Recall that
this was the area that received astronomy and mathematics. The Babylonians were adept at predictive omen astrology, dream interpretation
and fortune telling by casting arrows, and it is inevitable that the Jewish

---

48 The creatures had 'wings joined one to another' which seems to set them outside
astronomy and introduce the symbolism of spiritual life. Ezekiel was the first visionary to produce a four-faced god.

exiles came into contact with the astrologers of Babylon as they tried to sort out the zodiac (a zodiac that possibly included new categories of cardinal, fixed and mutable signs).

In fact, we know that the Jews did come in contact with the astrologers because both the books of Isaiah and of Deuteronomy roundly condemn astrology. Deuteronomy even says that Jews who take on astrology will be stoned to death. Yet generations later, after the freeing of the captives between 520 and 515 BC, some of the Jews (those who remained in Babylon presumably) were described by pagan writers as 'astronomers'. It seems that some at least had been 'infected'.[49]

The other evidence for an astrological/astronomical reading of the Ezekiel visions in Babylon[50] is the parallel development of the ephemeris: the tables of planetary positions that enable prediction. It was only in the eighth century BC that eclipses had started to be recorded in this part of the world. By Ezekiel's time, astrologers were able to calculate the conjunction of the sun and moon on a monthly basis, and of the moon with the other known planets, plus the eclipses. We know this because a tablet in cuneiform script was found dated 523 BC that gives this data.[51]

---

49   This blend of astrology and tradition had a peak in the 7th century AD when we find the Jewish population in an official role as court astrologers in Baghdad! One theory even states that the 'real Sabians' were the descendants of the Jews who stayed in Babylon after the Captivity ended.

50   There are many and detailed scholarly interpretations of this wild vision of Ezekiel. One of the most interesting is given in Jacques M. Chevalier, 'A Postmodern Revelation – signs of Astrology and the Apocalypse', University of Toronto Press, 1997. Chevalier examines many of the theories put forward that claim Ezekiel's vision as purely astronomical, being a vision of the northern circumpolar bear surrounded by the signs of the four seasons. As astronomy, the wheels within wheels represent the various astronomical 'wheels': the ecliptic, the equator, the two celestial colures and the band of the zodiac. This still presupposes an advanced astronomy in Babylon and it fails to explain why it was only the four fixed signs of what we know as zodiac imagery that were 'revealed' to the prophet.

51   The earliest known horoscope using the zodiac was also discovered on a clay tablet. It has been dated, from the longitudes provided, at 29th April 410 BC. Thus after 3000 years of sky observation, a chart was developed and this a mere 100 years before the conquests of Alexander injected Greek attitudes into the region.

The Rise of the Bureaucrats

In contrast to the visionary mind of Ezekiel, the lawyer Deuteronomists of the Babylonian Captivity no doubt looked upon the astrologers of Babylon and saw only 'star-worshippers' and 'fortune tellers' because that is all the mind without Vision is capable of – and they were punishers.

'If there be found among you, within any of thy gates
which the Lord thy God giveth thee, man or woman,
that hath wrought wickedness in the sight of the Lord thy God,
in transgressing his covenant

'And hath gone and served other gods, and worshipped them,
either the sun, or moon, or any of the host of heaven,
which I have not commanded;
'Then shalt thou bring forth that man or that woman,
which have committed that wicked thing, unto thy gates,
even that man or that woman,
and shalt stone them with stones, till they die.'

– Deuteronomy 17:3,4,6

Of course, the Deuteronomists were among those desperate to hold onto their identity: when any dominant religious or political group suddenly loses its power base and becomes a minority this is inevitable and involves a setting out and tightening of the rules of identity (which implies that someone seizes the authority by which to define the rules), and they will ban any uncertainties and other points of view. The *formalities* associated with identity come into ascendancy, while the *spirit* that breathes life into the tradition is denied. From the point of view of those setting themselves up to enforce a consistent framework for identity preservation, any Holy Spirit breathing lines of life across the Glory would be too unpredictable, too apt to 'blow where it will', to be relied upon when it is a question of survival.

This is what happened in Babylon and its chief agents were the lawyer Deuteronomists who strove to transform the emphasis of the holy writ, from the Glory to the Law. So Moses and his tablets of laws became central to biblical organization and the visionary features of the First Temple were expelled.

As the period of exile lengthened, this view of Jewish tradition set in and many former prophets lost their official place in the canon. A sort of blindness to holiness resulted. No more were many features of First Temple thinking allowed: there was no longer a heavenly host of angels in a visionary battle with evil, no more a king with a holy nimbus indicating divine authority.[52] The Deuteronomists reinterpreted the heavenly host as the exiles themselves destined for a literal holy war with literal enemies, and they blamed the corruption of the kings for the fall of Jerusalem. It all became material and lost its guiding myth and invisible, feminine Wisdom figure of the Temple.

Bureaucratised, the exiles finally returned to Jerusalem under the new High Priest Ezra, where they laid the foundations of the Second Temple.

Saving Enoch

Meanwhile, those who had been allowed to remain behind in Jerusalem during the absence of the exiles had continued their tradition of visionary theology graced by the transfiguring feminine Presence; these were shocked by the emptiness of those who returned, backed by Persian

---

52 The sixth century BC also brought about a general demotion of ideas of kingship and a corresponding rise of (relative) democratic ideas. Rome had her last king and was declared a republic; in Athens, Solon's laws legislated that there be a 'popular vote' by 400 representatives from the low third class; in India, Jina was the first leader to rebel against the entrenched caste system; and we have public libraries for the first time – a truly democratic move applied to the ownership of knowledge. Perhaps it was the cyclical demotion of an authority that had lost its way that necessitated a renewal of spirit via the incarnation of so many powerful religious figures.

money,[53] to rebuild and take over the Temple. An inevitable consequence was that *they resolved to maintain their own version of Jewish tradition outside the 'corrupted' Second Temple.*

They and their descendants were the ones who wrote down the Book of Enoch, based on the old oral tradition and which text also condemns the nature of the Second Temple priesthood. The Book of Enoch celebrates Prophet Enoch's ascent into the angelic realm and his vision of heaven, where among other things the secrets of astronomy are revealed to him. *This is the 'prophet' claimed by the Sabians*, the prophet of the book of protest against the totalitarian lawyers and their dead religion. In the Old Testament, Enoch is said to be the grandfather of Noah, and like Noah, those who had remained in Jerusalem, built a metaphorical ark and took their knowledge far out to sea....

In Babylon, we have the legacy of a healthy cross-pollination with astrologer/astronomers bumping up against visionary Judaism, perhaps absorbing much of its symbolic and story-filled approach to wisdom that may in turn have helped finalise which pictures in the sky became our zodiac. On the other hand, many Jews who lost their vision and may have been at a loss to express what had gone out of their hearts drifted toward the Babylonian learning, its megalithic heritage and struggle towards a perfect astrology.

From the point of view of sacred history, the real question is whether or not the Babylonians absorbed in part the Glory being abandoned by the Jews of the Babylonian Captivity. I like to think that they did; I like to think that the zodiac itself came alive, that it became a revealed knowledge, in anticipation of Pythagoras and the Western canon.

These were the generations of revelation generally and more than one received according to his capacity and the grace of the Holy Spirit. It seems that the zodiac was one such revealed knowledge.

---

53   Babylon fell to the Persians under Cyrus in 539 BC. The last ruler in Babylon, Nabonidus, had been told in a dream to rebuild the Temple of the Moon God Sin, after ten years of political instability. The King said that his subjects had offended the Moon God. Persian rule of Mesopotamia continued until 331 BC.

Removing the Traces

In the new Jerusalem of the late 6th century BC, the First Temple group became convinced that the divine breathe or Spirit-that-blows-where-it-will was no longer present in the new, Second Temple.

> 'There is no remembrance of former things;
> neither shall there be any remembrance of things that are to come
> with those that shall come after.'
>
> *– Ecclesiastes 1:11*

This is telling us that we will not remember our former visionary traditions and that we will lose the ability to remember 'things that are to come' – to prophecy on the divine potential of the soul. The law and literal approaches to knowledge had displaced the living wisdom and the result was that gnosis, the religious sight that the ancient figure of Wisdom[54] represented to them, was effectively banned among their society – and so the folk of the First Temple removed themselves.

The exiles in Babylon lost – were encouraged to lose – their insight, their direct 'sight'. This insight, the ability to see celestial substructures and mythical story images, is the feature that keeps a tradition alive. Those who don't have the gift – the lawyers – are always suspicious of those who do.

This is always happening; it is happening today. It is the duty of practitioners to seek to disclose this sight to themselves and within themselves. There is a spirit that inhabits any true revelation, and this spirit must be fed if a tradition is to live.

It should be clear now that, at this point in our story, history becomes

---

54   The Jewish Sophia or Wisdom has existed under many names recorded over the past 3000 years: Asherah, Ashtareth, Astarte, Lady of the Sea, Isis (whose temple was oriented towards the bright start Sirius). Perhaps the most familiar to us is the Babylonian Goddess Ishtar, who was both Earth Mother and Evening Star and The Queen of Heaven.

no longer a question of place at all; *it becomes a question of whether knowledge is dead or alive,* wherever it is taught. It is this Spirit, this breath that is Wisdom, Holy Spirit, Shekinah – figure, movement, presence – which we must follow in our astrological journey. She will lead us now beyond time, to one of the ancient cultic myths of the 'former things'.

This is the story of the Fallen Angels – the angels that brought astrology to humankind!

THE FALLEN ANGELS
Chapter Four

My mouth shall speak of wisdom; and the
meditation of my heart shall be of understanding.
I will incline mine ear to a parable;
I will open my dark saying upon a harp.

– Psalm 49: 3,4

Divinatory art... can always, to the degree in which it is interested, attract
insidious interferences. In other words, many cannot remove the veil of his
ignorance except by or through something which transcends his individual
will; for the individual curiosity, all 'oracle' remains equivocal and may
even reinforce the error which constitutes the fatal trap of such destiny.

– Titus Burckhardt

Original Myth

We have seen that, in the time of the Babylonian captivity, both the
Second Isaiah and the 'Deuteronomists' were concerned to alter the old
ways and to expunge certain myths that held and reflected the mind of
the First Temple. These went underground and, during the period of the
Second Temple, the original myths were passed along from generation to
generation. Several such stories tell of the coming of the Fallen Angels.

It will surprise nobody to learn that one major source for the story of
the Fallen Angels is The Book of Enoch, the aforementioned touchstone
for the world view of the Sabians of the ninth century AD. Harran, the

Vision text written down as the disenchanted wisdom line, went underground.[55] Yet the various sections that comprise Enoch were uncovered only 200 years ago in Ethiopia[56] and in the 20th century at Qumran. Prior to this, the Book of Enoch had been lost to the churches for at least 1500 years. Why? One strong reason for its disappearance is that Enoch was condemned by the formal church hierarchy as part of their war with pagan astrology and with 'gnostic' Christianity in the first two centuries of our era. (See chapter 5: The Knot of Christos.)

For our present purpose, which is to discover the history and development of the mental attitude which saw astrology condemned, trivialised and finally ignored as a public discourse, there are many clues in Enoch.

The story of the Fallen Angels in Enoch combines two interwoven myths.[57] One thread of the story tells how the angel Asael (represented as a fallen star in another section of Enoch) rebelled against God and brought several branches of God's sacred or hidden knowledge to the Earth. The second story tells how 200 rebel angels bound themselves by an oath and descended to Earth to marry human wives. The demon offspring of these unions were giants to whom the leader of the 200, Semjaza, taught the forbidden knowledge. The giants proceeded to kill the original inhabitants of the Earth. Furthermore,

'... they began to sin against birds and beasts and reptiles and fish. Then the earth laid accusation against the lawless ones.'

– Enoch VII: 5,6[58]

---

55   The current formal dating for Enoch sets it as early as the third century BC but the oral tradition and certainly some of the written records go back to at least the time of the Old Testament under discussion.

56   The Ethiopian find was made by a Sagittarian freemason named James Bruce, who may have known what he was looking for.

57   I am indebted to Margaret Barker for her thorough analysis of these stories and their possible meaning. See Barker: The Lost Prophet: the Book of Enoch and its Influence on Christianity, SPCK/Abingdon Press 1988.

58   The Book of Enoch, translated by R.H. Charles, SPCK 1997. All quotes are from this edition.

This is a myth about hybrid demons who committed sins against nature – environmental crimes.

The branches of sacred knowledge brought to Earth by the Fallen Angels were metallurgy, medicine (the 'cutting of roots' and acquaintanceship with plants), writing, cosmetics and astrology. Astrology appears in every list by scholars of the forbidden knowledge (or 'eternal secrets' as it is expressed in Enoch); and here we must recall that the term 'astrology' indicated a way of thinking that took in astronomy, mathematics, myth and magic. As well, astrology was linked with the setting of the calendar and the correct times for rituals that were considered doorways between heaven and earth.

Why Astrology?

It has to be asked – why astrology? How can astrology be powerful enough to be set as a forbidden knowledge? The question arises, of course, in the context of the contemporary trivializing of the subject, in which astrology as a public discourse is represented by sun-sign fortune telling paragraphs in the Sunday press, and where astrologers are embarrassed to admit their practice to any intelligent member of the public.[59]

However, as our history thus far has been at pains to point out, *any* knowledge is either alive – transfigured – or it is dead. It is a question of whether or not the blind lawyers have got hold of it from their position as the drivers of history and owners of the records. Metallurgy, medicine, writing, cosmetics and astrology are all what could be termed 'revolving door' bodies of knowledge; they can spin down, becoming dead or 'fallen' or they can spin centrifugally to the light, and retain their beauty and sacred office.

Take cosmetics, a seemingly trivial gift of the angels. The application of cosmetics can either enhance or conceal and it is this dual possibility

---

59   … much as, so far as the author's experience in concerned, many public intellectuals attend astrologers 'on the quiet'!

which indicates that the term cosmetic may imply a greater mystery. The Fourth Way teacher, the late Dr Phillip Groves remarked that the task of the human mind at its best is to enhance the gifts of God in nature. We are given the raw materials but we must discover their use and creatively add to this use for the good of mankind. Our enhancement brings nature to fulfilment. On the other hand, we can choose to disguise a corrupt motive, to steal an idea and pretend that it is our own and to imitate that which we are not, for our own ends. All this is 'cosmetic'.

Likewise, metallurgy, medicine and writing are dual systems and will have consequences – intended and unintended – when stripped of their living quality. The difference between building a bridge and manufacturing armaments has such a force of contrasting uses and intentions at back of them that it is unnecessary to comment any further.

Prophecy to Prediction

During the period of the Babylonian Captivity when the second Isaiah was making his contribution to the canon of the Old Testament, he used the term 'the former things' frequently, referring to the older mythology.

Isaiah writes:

'I have declared the former things from the beginning;
and they went forth out of my mouth, and I shewed them;
I did them suddenly, and *they came to pass*.'
– *Isaiah 48: 3* (emphasis added)

Here and elsewhere, the Second Isaiah is suggesting that the older myths were in reality *predictions* and that, as these predictions had been fulfilled in the captivity, such descriptions could be put aside. They could be dropped from the religious canon.

However, prophetic myths of power are not historical predictions: prophetic myths are revelations from the world beyond history, the timeless realm, and as such are *always happening*.

The confusion between the terms 'prophecy' and 'prediction' probably began during the period under discussion, as this was a period when there was the development of a simple ephemeris[60], which provided astrologers with the ability to anticipate some planetary conjunctions and it was also the time that writings in the form of prophetic vision were being banished from the records.

The slippage from a prophetic language to one of prediction was sealed when Alexander conquered Mesopotamia in 331 BC, which brought the Greek culture to ascendency across the entire region. Alexander had been a pupil of Aristotle. This is important as it was Aristotle who abandoned the living line of wisdom as taught by Pythagoras and in a limited version by Plato, who had already put aside the tougher side of Pythagoras' teaching.[61]

The Greek conquerors soon began to examine the knowledge of the conquered Chaldean Priesthood. They extracted the zodiac and further developed the ephemeris until it became a mathematical tool capable of calculating (inaccurate) detailed horoscopes. This process was overhauled and completed by Ptolemy in the second century AD, who conformed to the Aristotelean attitudes and believed astrology to be strictly cause and effect and to contain nothing in particular for character and spiritual development. This became the predictions-for-the-body-in-space basis for the future of Western astrology. In the process, the Greeks projected back onto the Chaldean Priesthood a mythical notion that they had been the custodians of a secret knowledge, now recovered *and to be improved upon* by the Greeks. The Greeks streamlined and exported their version of astrology (and much else).

Of course, on the surface, public Greek culture – with its myths, gods, academies and theatres – would appear to be compatible with and

---

60   Recall that there was nothing like this in Europe.

61   Aristotle left Plato's Academy in Athens and went on to assert that any universal realities (or spiritual archetypes) could only be accessed intellectually by abstracting general characteristics from material things. It was Aristotelianism that the West inherited via the new universities of the Medieval period.

easily integrated to a culture housing prophetic revelation but this is not the case. The cool, fatalistic pronouncements of the Greek gods and their Aristotelian philosophy and highly rational[62] search for a knowledge which operates within systems are at variance with the poetic *energies* of prophetic utterance. Dionysian chaos is only distantly related to the true divine fire, which is the seat of prophecy.

So what is prophecy and how does it differ from prediction?

When we tell an event in a mixture of symbol and imagery, we enter a part of collective Mind where literal truth and creative lies are strangely mixed, in which an image is presented that is really a source-seed for the gradual unfolding of meaning. Such an image speaks to the soul, directly and without the structure of reason. The source images connect within the flow of a visionary story. Speaking or thinking in this manner is all but lost as a function of communication in the West, except in the world of certain kinds of artists.

Prediction, on the other hand, will always bring the mind back into history and therefore away from spiritual striving: in fact, the mind is *attracted* to prediction.[63] The Greeks of those times – as abstractors, as intellectuals – were bound to take astrology down the road of intellectual fancies; they were like 'bright boys' who recoil from the messages from the Dark.[64] Yet the other potential of astrology remained, always there in the shadows.

The unintended consequence of the coming of predictive horoscopy was a diminishing of our ability to read astrology in layers as a prophetic language or, rather the ability to *argue the validity of this poetic approach to knowledge* and to decision making in our culture. We are living with the legacy; for by adopting 'prediction' as their mandate, astrologers became vulnerable as a group, both physically – as

---

62   The mentality expressed as ancient Greek culture is the only historical attitude comparable to the scientific emphasis that emerged in the late Renaissance.

63   Attraction is a genuine hindrance to the Path.

64   See the chapter on Pluto.

no powerful force will tolerate a group of people claiming to know the future already – and as a tradition that was really intended to preserve the prophetic mind. Astrologers did not realise how vulnerable they had become until the late Renaissance and the coming of hard science, with its 'prove it' and humanist control-of-nature forward march. By then, astrologers had lost their credibility and their mandate as defenders of the sacred dimension in humanity, and so astrology began its slide into mindless 'popularity'.

The story of the Fallen Angels is straight from the Vision tradition and its prophetic language. It is a description of the means by which the creation becomes – and is always becoming – corrupted. Anyone who considers the Fallen Angels will not only come to understand the structure of corruption but may also recognise an inferred path to redemption.

Prophetic stories are always happening.

Corruption or Ferment?

There is a story from the Ismaili Gnosis (a branch of Shi-ite Islam) which also references a Fallen Angel. The Ismailis contend that, at the beginning of time, one of the angels was tardy in recognising Divine Unity and that this tardiness created a *separation from God* and caused the angel to fall.[65] The hesitation on the part of the angel was the first manifestation of 'blindness' and blindness *is* the fall. Wisdom is lost to us. As Proverbs tells us:

'Happy is the man that findeth wisdom, and the man that getteth understanding.

'The Lord by wisdom hath founded the earth;
and by understanding he established the heavens.

---

65   The medieval mystic Suhrawardi taught (in an insight of pure spiritual beauty) that the Angel Gabriel possesses one wing that is day and one of night, as a living emblem of this moment of separation. In the astrology of the medieval mystic Ibn Arabi, it is the angel of the sign Capricorn that holds the keys of day and night.

'My son, let not them depart from thine eyes:
keep sound wisdom and discretion:
'So shall they be life unto thy soul, and grace to thy neck.'

— Proverbs 3: 13,19,21,22

Without Wisdom, nature and all knowledge are dead. Margaret Barker claims that such blindness leads also to the abuse of nature and of women and this author is inclined to agree with her with one reservation, which is that our blindness does not kill the thing itself: the glory still exists. So the question becomes 'how can we recover our sight?'

Therefore, again – why astrology? Why was it considered to be a sacred knowledge, powerful enough to be stolen? Why is it associated with a visionary text, suppressed by the incoming hierarchy? Moreover, what is 'dual' about it – how can it rise? Put simply, if astrology received an infusion of the Divine Wisdom and returned to life, what would our experience of astrology become?

Here is a proposition. If the stories of the Fallen Angels are source myths about the coming of the slippery arts and their corrupting operation, perhaps we need also to look at the term *corruption* in a fresh light. If we add some spiritual sight to blind corruption, surely it will transform corruption into a kind of ferment. Ferment is 'living' corruption, a darkness that is alive also with the possibility of regeneration and growth. Is it possible that the Fallen Angels have a living purpose, as ferment that promotes growth? If so, the question must arise 'what are we to grow?'

All the branches of knowledge brought to earth by the Fallen Angels are functions of divine being when viewed in the integrated manner of the ancient mind. Metallurgy is enlivened by alchemy, medicine by exorcism.[66] Astrology too appears to be about something 'on the edge of consciousness', due to its language of pattern, light and vastness; and it

---

66   Margaret Barker remarks upon the 'mutual impoverishment' that has resulted from the separation of medicine and religion.

was astrology that carried the knowledge of cycles, of what astrologers call transits.

What if astrology (as then understood by the Visionaries) marked out the cycles that contain the possibility of 'the maturing in each soul *of an aptitude for the angelic state'.*[67]

What if we are required to grow a soul?

---

[67] Henry Corbin: 'Temple and Contemplation' Kegan Paul International 1986, p. 163 (emphasis added).

THE ANGEL BODY
Chapter Five

Wisdom went forth to make her dwelling among the children of men,
And found no dwelling place:
Wisdom returned to her place
And took her seat among the angels.

– The Book of Enoch: XLII, 2

The Shifting of Blame

There was another massive shift that took place in Babylon as a result of the deletes and the new emphasis in the law: this was a shifting of blame, and no-where is this more evident than in the subtle manipulation – and then insertion – of the core 'Judeo-Christian' story of Adam and Eve.

'Judeo-Christian' is here placed in inverted commas because the story of Adam and Eve is never mentioned in the Old Testament except in Genesis and Jesus does not refer to it in the New Testament, which seems odd if we consider the claims by Christians that Jesus Christ came to fulfil the prophecies of the Old Testament, and that he quoted scriptural passages from the Old Testament at moments of fulfilment.

The Adam and Eve story was very possibly added to Genesis by the Deuteronomist bureaucracy as the Old Testament was put together during the exile, because *the tree and the snake, so important in Genesis, were symbols of the Divine Wisdom and her Glory.* Genesis as we know it may be another distortion of 'the former things' and the re-telling of Genesis simply a polemic against Wisdom.

In the well-known story of Adam, Eve and the Snake, it will be recalled that the One God tells Adam and Eve to *stay away from that tree!* (of Wisdom), while the serpent advises Eve differently. The serpent tells Eve that she *ought* to eat of the fruit of the tree because, if she does, she will become 'like God' – that is, her spiritual vision will be opened if she associates with the Divine Wisdom! Yet after the Second Temple 're-writes', there is a change of emphasis: now it is Eve and her eating of the apple that is the cause of the corrupting of the world. 'Adam and Eve' becomes a major exercise in removing the charge of corruption – and the examination of the nature of corruption – from the Fallen Angels (who could then be said not to exist) and placing it in the hands of the archetypal woman, Eve, and the Serpent of Wisdom.

The consequences of the entry of the Adam and Eve story into the Western world cannot be underestimated. At its worst, the trashing of the living feminine principle has led steadily to both the anathematizing of the woman's point of view and the trashing of Mother Nature, the manifestation of sacred Life.

A further result of the Deuteronomist takeover of theology and its export to Jerusalem was that the Second Temple Jews isolated themselves in their region of the world. Prior to the captivity, there were many gods in Canaan: now there was one, newly minted 'jealous god' who was a punisher and who would tolerate no rivals, whether graven images or rituals in natural groves or star-angels. How did they arrive at this belief in the punisher-god? The problem of the Second Temple thinkers appears to have been their dead understanding of the very term 'One'; for bureaucrats will always interpret 'one' as a numeral and so as One God, while those who love Wisdom would know that One signifies also a quality – and that quality is Unity. This is unity-in-diversity, the plane of life underlying infinite forms.

As the Bhagavad Gita so beautifully states, whatever road you take 'you come to Me'.

Ideas are powerful. Judeo-Christianity spread, as we know, and today in the Western world – and this despite the fashionable atheism and the fall in church attendances – the majority of people replying to the census still claim Christianity as their formal religion.[68] However, if religion is only blind religion, religious wars will result, as this kind of thinking leads to all kinds of fundamentalism (including market fundamentalism). This is why knowledge of the difference between the living and the dead manifestations of any scheme is vital. As astrologers, we too have a duty to make this separation and ponder its implications, particularly as we are trained to recognise living symbols within story and the many levels of interpretation possible in the symbolic.

Ideas are powerful. Let us go back to our story.

We begin again: at this moment – when the final form of the zodiac is beginning to be used in Babylon – the folk of the Vision Temple in Jerusalem are written out of history, remove themselves and go underground, taking with them their Fallen Angels, Wisdom and their own stars-as-angels. Later, the inheritors of this stream commence their writings which roundly condemn the incoming materialism.

> 'And these are they who judge the stars of heaven,
> and raise their hands against the Most High,
> and tread upon the earth and dwell upon it.
> And all their deeds manifest unrighteousness,
> *And their power rests upon their riches…*
>   – *The Book of Enoch: XLVI, 7* (emphasis added)

The Vision Temple saw the difference between those with money-sight and those whose riches remained 'in Heaven'.

Communities of Light

In addition to the story of the Fallen Angels and the rejection of the new religious vulgarity, the authors of Enoch also state that there exist people whom Wisdom *did not seek but found anyway* and that she 'dwelt with

---

68   In Australia, it is 61 percent.

them, as rain in a desert and dew on a thirsty land.'[69] Who are these people whom Wisdom loved?

Those blessed with religious sight tell us that there exists an invisible web of relationships, a totality much like a vast eco-system that takes in the human and spiritual worlds, and through which all things commune in love with all other things. There are some fine threads in this web, which connect various 'communities of light' through lines that link their separate revelations. These communities of light are people who evolve and find each other across spiritual space because they seek the inner or esoteric meaning of the symbols of their tradition and this brings them to its heart.[70] Without the static that infects the denser levels, they locate each other.

It is also said – and has been mentioned – that there exist cycles of *sacred* history in addition to mundane, recorded history. It is believed that we are living now in one of the prolonged cycles in which humanity loses its knowledge and its capacity to participate directly in the continuum that is God. We enter this cycle due to some kind of doubt or 'fall' and we slide away from spiritual certainty.

The Ismailis believe that the role of individual human beings is *to assist in the redemption of the Fallen Angels by one's own self-effort* towards renewed certainty. They believe that the entire (sacred) history of our world is an effort to restore the Fallen Angel to his former rank. However, as we are fallen – blind – we require mediating symbols to remind us who we really are, symbols that are a language that reaches beyond our confusion and our disappointment and speaks with the soul. Right symbols, properly treated, can bring us to a point where we begin to identify again with our Angel Body in potential, rather than with the rolling crises that mostly constitute our lives.

This brings us to the Angel Body itself.

---

69   Book of Enoch: op. cit. XLII, 3

70   Knowledge of these invisible threads is a mystery of Neptune in astrology.

The Angel Body

Every human being is born with a potential that is like a seed set in a secret compartment of the heart. The seed is constructed of pure qualities such as beauty and goodness. However, the seed needs water – not the poisonous spray that serves in much of contemporary culture – if it is to undergo its proper stages of development. It needs the living water that is bestowed when the heart loves something that is sacred to it, or serves the sacred dimension of that with which it comes in contact. This seed needs to sprout.

The process of sprouting the seed marks the start of the formation of the first level (or sac) of the Angel Body and (under the right conditions) will become the foundation for further development. However, these beginnings are slow work and it is to be emphasised that it is the effort to do the work that brings about the transformation.[71] Such transformation can also change permanently one's manner of operation in the world. Quite simply, we can become another person.

Like the glorious sap that is the earth, the Angel Body emanates from and surrounds the human being: it reaches inwards with love to the divine infusion of the pure archetypes, which it expresses.

The Angel Body needs to be encouraged to begin to stir. The first step is to discriminate between receiving poison and receiving living water and to choose the water or *healthy context*. Concurrent with this there needs to be a resolution to restrain negativity within oneself, including negative speech, and to let go of contexts that support such negativity. This is hard as we are creatures of habit but there comes a time when the adjustment may achieve a critical mass through practice. The seed suddenly sends up a shoot, responding to a fresh heat emanating from Mind.

Of course, all legitimate traditions legislate in favour of positive personal practices and there is no need to repeat here their well-founded

---

71   In the Christian tradition, this stage is called 'cleaning the cup'. For the Greeks, it was the Lesser Mysteries and for the Muslims, the jihad.

wisdom. What is being emphasised instead is that a *tangible* result may be achieved from efforts made to reverse and purify the habitual tendencies, which are reinforced by the automatic social habits of our everyday environment. Our efforts are like a perspiration that flows as a sap to the seed root. Critically, our efforts need to become *conscious* and properly directed[72] if the Body is to become permanent. It is no longer a question of simply 'having a good day': more is required.

In a lecture to his students in 1994, Dr Phillip Groves likened the construction of the soul to an alloy in chemistry, where various components are melted by the heat of our efforts and forced into a new kind of fusion. Dr Groves stated that a *fixing agent is then required to seal the new being into place* so that we cannot lose that which we have built. This principle can be applied at the simple 'feet on the ground' level; for instance one may after a period of dietary reform introduce regular exercise into the life. The exercise routine acts as the seal so that the old habits have difficulty finding their way back into the mind.

The Angel Body levels each require a seal and the process follows the same fixing route but in a powerful manner, where the setting of these seals is equivalent to a quantum jump in physics: something is added and you are somewhere else, forever. There is a sheath present, absolute.

Interestingly, one effect of the arising of the little heart-shoot will be that 'things get harder' as the new shoot strives to take root and move upwards to the light. The first soul-sac may be formed and sealed into place in order for further growth to become possible but it is certain that the old self will kick as one tries to go on. Yet this is the beginning of the lasting Angel Body, which all legitimate traditions have as their human end and for which construction their practices were created. [73]

72 The Fourth Way teaches that improperly directed or mechanistic efforts can also result in permanent change but in this case it is damage.

73 This parallels the construction of that which other traditions call the Astral Body and the Resurrection Body; in the Fourth Way School the commencement of the body is called the Kesdjen Body.

The Chart as Sanctuary

We now enter the country where our inner-plane astrology resides, for the astrological map is an ideal setting for the growth of the first sheath of the Angel, if one is aware of such a possibility. Astrology supplies a wonderful foundation for the re-ordering of the mind that is a precondition for change. Contemplation of the astrological map and its planetary functions, travelling as rhythmical storytelling around a cycle of qualities, will begin to open the necessary pathways through which light may be received and reflected. If one regularly uses the language (which means contemplating other astrological maps as well as returning frequently to one's own map) the illumination will intensify.

The chart will be revealed at last as *a sanctuary for one's own Angel Body in potential*.

In fact, stars and angels often appear as interchangeable terms in scripture; and there is an aura about this language that transcends the easy explanation of metaphor.

> 'And I saw other lightnings and the stars of heaven,
> and I saw how He called them all by their names
> and they hearkened unto Him.'
> — *ENOCH XLIII : Astronomical Secrets*

In the sky world as pondered by the religious mind, it could be said that the star is the temple-body of an angel, with its language of light. The earthly correspondence is the human being with his or her astrological map into which the potential angel is invited to make its home. Our own efforts are the invitation and the astrological map is the fertilizing container for the development of the first sac of the soul.[74]

Our transits are of immense importance in opening the opportunities for Angel Body development: those moments in time when a planet

---

74   Refer chapter 5 for speculation about the other soul-sacs.

passes over or relates significantly to a planetary or angular feature in one's map. For example, the angel of Uranus grants the intellectual courage to become a stranger to common habit. Neptune encourages the emptiness of the heart that enables the angel to be seen. Astrologers will surely recognise in the seal the voice of the Mysteries of Saturn. Saturn transits make a fine supporter of fixing practices; and the transits of Pluto assist greatly in helping us to recognise when and where efforts have to be made if the Angel Body is to be constructed. Pluto is the planet of will and right effort. We will speak more on this later.

Yet strangely, as effort commences, the existence of the map and its influence forces a confrontation between the intuited spiritual potential of the individual and the confinement of map tendencies. The birth chart may appear as *both a limiting and a liberating gift*. Transits too may be perceived to either limit being or call forth the soul. It is a struggle, first to learn to respond to the planetary call and later to learn to harness the energy.

Remember: astrology is a revolving door art.

The Eye of the Angel

The sealing of the first sheath will see some results or fruits of the path. One fruit is a greatly increased capacity to perceive the inner plane of beauty underlying everything on earth. Even one opening of one eye of the angel is like a glimpse of Paradise, an open sesame into a chamber full of light – before the door gently closes again returning the hungry human to his or her struggles. We have glimpsed what it *may* be like to see with the eyes of God and hear with the ears of God, without separation.

Moreover, the brief opening of the eye is also a cry to a star-angel, through whose mysterious help the possibility of a more refined set of results from the dance of transits arises. The ancients recognised the life and the stories that are the star-angels. The Sabians knew the angels that were and are our planets, in our little galaxy here – silent, beautiful – waiting for us to obtain enough inner light to see them. The link between the ancient communities of light and modern practice is located here and if recognition of the timeless, esoteric purpose of the map takes place, real effort will commence for the astrologer. Real recognition – in the gut, as they say – *is* real effort, a resolve and beyond theory.

It is a pity that we do not have more cultural support for the teaching of astrology as a Wisdom School; and it is a pity that so much knowledge has been 'gutted', hence destroyed: first by denying its life and so moving it onto historical territory where it can be patented, then by subjecting it to a 'rational' critique by the new hierarchy through which it can be mocked and marginalised. Finally, we ignore it. This is the real 'death of God'. God is not dead, although the patriarchal, anthropomorphic version of Him may be; and here those with spiritual sight have to be damned careful because, if we allow the death of the historically constructed One God, we may end up with a vacuum to be filled by the hybrid Fallen Angels of our own times: fundamentalists, capitalists, cynics and materialist science-worshippers (fill in your own examples).

As stated, true prophecies are always happening.

What really matters is the objective state of our practice, for unless practitioners allow – even in their secret hearts – that astrology can become again a temple-threshold and a pathway to the re-formation of the Angel Body, it is indeed worthless. It is the work of astrology to help recover the Angel Body, and its mode of perception that discloses the earth as sacred and made of divine symbols. As we live in a time when the restoration of such spiritual sight is desperately needed, the state of one's practice is a trust.

And so, in the place and in the time of the rising of the separate One God, the 'other knowledge' vanished. By about 100 BC, the Vision theology and Enoch had found their way to Qumran and to the mystical Essenes, while formal history Hellenised the region. It was becoming clear to anyone with eyes to see that the time was ripening for a new teacher of Wisdom to appear and for the eyes of the Angel Body to be renewed, and as Jubilees approached,[75] such was anticipated.

This is the subject of the following chapter.

---

75   Jubilees is the great cycle when all debts are cancelled.

THE KNOT OF CHRISTOS
Chapter Six

'Yea, before the sun and the signs were created,
Before the stars of the heaven were made,
His name was named before the Lord of Spirits.'

– ENOCH: XLVIII,3

By faith Enoch was translated that he should not see death;
and was not found, because God had translated him:
for before is translation he had this testimony, that he pleased God.

– Hebrews 11:5

Rome and Jerusalem

The period from the revelation of the zodiac in Babylon up until about 100 AD saw the spread of astrology into the Greek and Roman worlds and the steady rise of belief in an astrological determinism across all classes. By the time of the Roman Empire[76] under Emperor Augustus (and the time of Jesus Christ), astrology was a powerful contender for the control of ideas at the official level.

This was not the astrology of the Angel Body. How could it be under a new Roman empire with an emperor who doubled as a religious cult figure and who was well received as such in the Eastern part of the Empire – with the exception of the Second Temple Jews who set their

---

76   The Roman Empire extended from 65 BC to 363 AD.

One True God against all other gods and emperors?[77] No, this had to be fatalistic, predictive astrology – inheritors of the ancient divination practices, casting of arrows, tea leaves and the like, pulled together by Greek appropriations and mathematics – working within the Greco-Roman framework of the times. That there was a parallel line of development also flowering within the zodiac cosmology is indicated by the overlapping waves of esoteric thought, such as Hermeticism and Neo-Platonism, that broke over the first three centuries, all of which took 'As Above, So Below' as central to their philosophy.

However, esoteric astrology did find one tentative home among certain of the Roman intellectuals. Under Emperor Augustus, the architect Vitruvius wrote a treatise in which the qualities inherent in number are applied to proportion in architecture and where number becomes the cement holding together eight core 'Vitruvian subjects',[78] which included astrology. So astrology was acknowledged as a legitimate subject and touchstone in intellectual circles (as well as among some officials) in the first century.

Meanwhile, in Jerusalem and surrounds the longing of many for the open return of Lady Wisdom and for a *true* line of kings – those whose legitimacy is visibly expressed by the presence of hvarna or glory[79] – persisted and which as a movement possessed its own library and inner circle. Thus, when Jesus began his ministry, he was expected – or rather, a true King who would restore the First Temple was expected. However, the mission of Jesus was not to become a literal King in the Roman Empire.

So much has been written about Jesus over the centuries that this author is embarrassed to have to go there. However, Jesus is one of the

---

77   The Second Temple was finally razed by the Romans in 70 AD.

78   Vitruvius is raised here as his writing became the foundation for the great period in neoclassical architecture in Renaissance Europe and his 'living' mathematics influenced seminal figures of this period. We will return to him.

79   The Hvarna as the heavenly mandate that is bestowed and can as easily be withdrawn. In iconography, this feminine glory is expressed as the halo or nimbus.

key figures arising from the living Wisdom tradition and, as our Western world is still dominated by 'religious' Christianity (now mirrored by a rise in fundamentalist atheism[80]), Jesus is impossible to avoid if we are to break open the literal approach to knowledge that has become our unfortunate inheritance. In this project, it is hoped that astrological practice may be restored as one method of obtaining Wisdom's grace.

Therefore – Jesus 'the Christ'. It must strike anyone encountering Christianity that so little is known about Jesus apart from the stories in the gospels. The Roman historian Tacitus records that Jesus was crucified and the Jewish historian Josephus also mentions Jesus – and that is the sum of it from the writings of those times. How strange: surely some larger record about this travelling healer and wise man would have been made but no. The French scholar, the late Réne Guénon has written[81] that almost everything touching upon the origins and earliest years of Christianity is 'shrouded in mystery'. Why? Guenon informs us that one sector of the Islamic world considers early Christianity not to have been a general path for communities at all but rather to have been *an initiatic way.*

What does he mean– initiation into what?

Now significantly, the third (and the second to a certain extent) sheathes of the Angel Body creates a man whose real activities are invisible to those participating in ordinary history. This can be illustrated clearly by a traditional cross, where the vertical arm signifies Heaven or the path of ever refining evolution toward the Angel Body, and the horizontal represents Earth or the common experience of historical man. It is obvious that the path of any initiate must first traverse the horizontal axis (in stages) until he or she reaches the central point, whence the vertical axis is opened to the initiate's sight.[82] Should the candidate then

---

80 The two fundamentalisms are bound to arise at the same time as fundamentalists need a projected 'enemy' through which to establish their own identity.

81 René Guénon 'Insights into Christian Esotericism', translated by Henry C. Fohr, Sophia Perennis, Ghent NY 1943.

82 The reading of the centre is the culmination of what is known as The Lesser Mysteries.

begin to 'rise' even one degree along the vertical, he or she will become invisible to those labouring along the horizontal axis – which means to the vast majority of us.

If the mission of Jesus arose from and was underpinned by the construction of all three sheaths of the Angel Body, it is certain that 'history' would not have been capable of seeing the beginning of Christianity clearly. The sheaths produce a beyond-time body– which means of course that it is also beyond any influence of our time-locked planets!

It is here that some brief speculation needs to be made about the three layers of the Angel Body, their possible effects and their astrology.

The Heart Body

Prior to any period of self-development, an individual will be tossed about by planetary influences, whether that individual knows it or not. The horoscope is a trap, with patterns that repeat and repeat under various disguises. Transits bring events to which the individual reacts and which he or she judges: 'this is good', 'this is bad'.

Sooner or later, certain capable individuals may become aware of their patterns and want to change. However, as stated previously, the first level cannot be approached without repeated attention to one's chaos and a parallel resolve to transform the character, supported by a fresh context and a clear aim. This is the time of Know Thyself. At a critical point in this effort to become conscious, it will become possible to achieve the first level (or Heart Body)[83] of the Angel Body.

It appears that the sealing of the first sac greatly modifies the individual's response to astrological events. The results from astrology become more refined. Events provoked by transits begin to be interpreted as

83   The author will be referring to the three soul sheaths as the Heart Body, Mission Body and Praise Body based upon perceived functions. These are not common terms. The Fourth Way School indicates three bodies that are both the fruits and enablers of action in the world: the first is the result of being of help to oneself, the second of being of help to mankind and the third of being of help to God.

qualities in and of time rather than as good-bad, impacting externals; and if this is seen to be so by the candidate, the idea may arise that one may be able to harness planetary energies and to co-operate with planetary timing. This harnessing is one practice associated with the first level.

Later, after the evolving Angel Body begins to acquire its seal, two key developments will inform the operation. Firstly, the mind acquires greater order and is freer from automatic connections and chaotic ramblings. Practically speaking, the mind changes the way it prefers to spend its time.

This new stability is the mind preparing itself to receive another kind of knowledge.

The other development has to do with time and consequence. The first sheath of the Angel Body appears to bring with it the capacity to receive back the consequences (or 'karma' if you like) arising from one's activities in the present. In fact, there is the creation of a rapid feedback loop, so that *any action brings its consequences in ever shortening cycles of time.* Critically, *the energy of the feedback provides fuel to further grow the Angel Body*, as one is compelled to see, accept and then utilise the results of one's oft-times poorly motivated or unconscious activities. It is a period of inner separation and striving.

The time between the first and second sheaths is sometimes isolating, as so many activities enjoyed by the masses cannot be tolerated at this time. For example, one cannot stare at a screen for long without disturbing the mind because it is screen and internet culture that are the dominant forms of the collective 'sleep' of our times.[84] Worse, it is also a time when the individual may unintentionally set off 'projections' onto him or her from others. If one says too much – sometimes if one says anything at all – a range of reactions follow from others who may, for example, scream at or become strangely obsessed by the initiate. It will become clear that this situation must be managed somehow (unless the candidate withdraws to a cave). At last it may occur that these shock

---

84   The Age of Aquarius has introduced the sleep of the World Wide Web.

interactions might be happening because *the additional energy that comes with projected activity is needed*. Yet such will caution the individual as well and encourage silence.

The process of learning to utilise the 'blow back' from one's intentions, conscious or otherwise, and the shocks following interactions with others (in which one may be found to be asleep or simply mystified) is a stage and a critical one. In such a process one becomes aware of the flow of energy and the flow and feedback loop that is one's own thinking. It may begin to be seen that thought itself can be harnessed and one's thought waves also utilised.[85] Thus we pass further into the 'interior' and begin to free ourselves from our own patterns and prejudices.

One fruit of this process is a departure from thought that is rooted in polarities. There is an arising of the possibility of a direct sight that sees without comparing.

And what of the body in history at this time?

The Heart Body period is *the* time when an individual will be required to take on tasks with an increased degree of difficulty and according to his or her evolving destiny. *The astrological map is still active but on a level of ever finer self-fulfilment*, a house being polished by transits and directions until it becomes a temple. It is essential in the preliminary work and the work from the first level to allow the planetary energies of Uranus, Neptune and Pluto (as Mind, Heart and Will) to *assist* one's tasks. It is the period of Know thy *Self*, beyond the Know Thyself of initial efforts. In fact, the combination of mental ordering aiming at Truth (Uranus) with the utilization (Pluto) of blowback (Neptune) and the completion of tasks (Saturn) creates the receptor for the next seal, that which enables the Second Body (the Mission Body) to appear.

The second seal is a spiritual organ; it gives now *the capacity to receive and transform the wounds of others* and to intervene where there is error. It should be clear that Jesus Christ had come into this second

---

85   This is a mystery of the rings of Saturn. Saturn's exaltation in Aquarius indicates its
      rulership of mental possibilities. Note too that Uranus has rings, although fewer.

seal and, as we shall see, there are others who possess a further sheath: the Mission Body.

The Mission Body

Anyone with half a mind must be aware that there exist in this world individuals with a greatness of being. These are the great teachers– those who speak directly to the dormant potential of the soul – or the great healers who intervene in a manner that seems simply to remove the disease. Then there are the great artists, for instance those composers who 'stand before God' as they strive (for example) to set a liturgy to music, or artists who paint pictures whose impact as beauty lasts and causes 'something' in the viewer to leap with recognition. These are the Level Two people. These are the teachers who 'only appear when the student is ready' and they are capable of being given a mission. The miracles of Jesus came from this level for, as G.I. Gurdjieff observed, a miracle is the intervention in one set of laws by a higher set of laws. The Second Level and its sealing organ enable such a man to absorb and transform diseases and 'karma' from the lower levels.

The Mission Body folk are those of a progressive invisibility to history. They have achieved identification with what Plato mapped as a divine body and which he sets *above* 'the heavens, the sun and the planets' in his scheme of rising mankind.[86] Thus as the refined levels can always see the grosser levels (this does not apply in reverse), these people may be able to utilise planetary forces on behalf of others or on behalf of the aim of an esoteric school, but *they themselves are free from such influences.*

It seems that Jesus spent much of his life on earth as one of the pro-phetic teachers and healers of the Second Level, and it makes spiritual

---

86 'Above' here is not literal hierarchy: it is more an illustration of progressive refine-ment that brings us closer to the Heart of Mystery. Such things are difficult to illus-trate.

sense to set his mission at an advanced level of this Mission Body, and to submit that his crucifixion and the flowering of resurrection stories set within a sort of mass confusion and 'shroud of mystery' marked a massive capacity to receive the diseases of others.

The passing over of Jesus was the passing into the Third Level, invisible to history. Such is the empty tomb.

The Praise Body

'Praise ye the Lord. Praise God in his sanctuary: praise him in the firmament of his power.

Praise him for his mighty acts: praise him according to his excellent greatness.

Praise him with the sound of the trumpet: praise him with the psaltery and harp.

Praise him with the tumbrel and dance: praise him with stringed instruments and organs.

Praise him upon the loud cymbals: praise him upon the high sounding cymbals.

Let every thing that hath breath praise the Lord. Praise ye the Lord.'
                                                                   – Psalm 150

What can be said about the Praise Body? Not much, surely, but there are some clues.[87] One clue is the continuous song of praise of the scriptural angels, those who praise the Glory of God and who never sleep. (In Enoch, 'the stars that never set'.)

From our own point of view, the summons to praise continually can only seem a little boring, as do representations of Heaven, and the saying 'the devil has the best lines' holds true in any assessment of that which makes life interesting and rich. However, this is the Praise Body and here the term cannot be as we understand it.

---

87   God loves clues.

It is said that the Bible can be read on four levels: the literal, allegorical, moral and the mystical[88] and while it may be too neat to suggest that this line of extended perception relates to the whispering from the sheaths of the Angel Body in potential – yet there may be a correspondence. If there is a correspondence between the Praise Body and a mystical reading of the Bible, how may we strive inwards to search it out?

All our traditions tell us that attention– presence or mindfulness– is both a practice and an end. That the angels of Praise 'never sleep' indicate that they are fully conscious; and so in a sense *the practice of mindfulness is also a prayer of the Praise Body.*

Perhaps the acquisition of the Praise Body and the synchronous disappearance from history may be like the coming of gratitude and the end of resentment, for we cannot be in a state of gratitude and hold resentment at the same time. *If one identifies a state or a quality (not an activity) that removes darkness instantly, it is connected with the Praise Body.*

Another clue is found in previous comments on the Heart Body, where rapid feedback appears to contract the timeline between cause and effect. Time qualities folds back on us psychically, much as spacetime is curved around matter. Thus by the time of the Praise Body, there is an ever-diminishing space-time folding in towards the heart and in the end, *this will collapse the heart into love and bring us into the presence of the Glory.*[89]

Here is an example of a no-time mystical reading of the teaching of Jesus about the necessity to forgive sins and 'turn the other cheek'.

Anyone who has experienced the illness and death of someone with whom he or she was formerly in conflict – a parent one has struggled

---

88   One of God's best jokes is to disguise the esoteric in the language we speak, which means that two people can hear the same statement in the same plain English and take away utterly disparate meaning, should one person have the Heart Body and one person not have it. This is the real Tower of Babel.

89   There is no way of speaking of these things without resort to spiritual markers such as Glory and Heart. It is hoped that the ordinary, practical examples will help us to navigate this mysterious and poetic place.

with, an ex-partner or sibling rival – will know that the time of life-and-death cancels all enmity. It is as if the enmity *never existed* or, as a friend once said, 'all the agendas no longer matter'. The formation of the Praise Body is the call to collapse time completely, so that this forgiveness, this 'it never happened' is able to occur at the precise moment of the transgression. Forgiveness is the Glory that is *built into* sin, always there– past, present, timeless.

We do not forgive each other: we only remember that forgiveness had already occurred and was always present.

This seems a possible Praise Body understanding of the teaching about sin. It moves, via the collapse of time, from the moral advice to forgive others their sins into the knowledge that sin and forgiveness are one and present.

There is also a clue about praise in the writing of Jacob Boehme, the Renaissance cobbler who received a revelation from the angelic world and wrote it down only after thirteen years of contemplating its meaning.

Boehme writes: 'Christ, according to the eternal Word of the Deity, eateth not the substance of heaven, as a creature, but of the human faith and earnest prayer, and the souls of men praising God are his food, which the eternal Word that became man eateth...'[90]

It appears that praise is an energy feeding the moving Spirit and that the Praise Body is a fulfilment of the desire of Spirit to be nourished.[91]

Eternal Rivalries

After the blast of mystery that created Christianity, there were three hundred years of intense struggle. The Roman Empire, always ready to crush subversives, maintained its program of summary execution of all perceived rivals to power. Early Christians were commonly

90   Jacob Boehme: Mysterium Magnum, The Seventieth chapter, quoted in Basarab Nicolescu 'Science, Meaning & Evolution, the Cosmology of Jacob Boehme', Parabola Books 1991.

91   Recall also the myth about resisting human hearts being eaten, chapter one.

killed. However, the suffering of the Christians caused them to become organised and to set in place a consistent hierarchy of bishops, priests and laity. Meanwhile, the Roman emperors took on the mantel and title of Sun King and this placed the Empire in a position from which it became capable of recognizing the Sun King Jesus. This they did and in 315 AD Rome converted to Christianity, merging two systems based upon the solar principle and its worldly hierarchy.

There was also intense rivalry between Christianity and astro-fatalism over how history would answer the question 'through what means does God manifest in the world?' Given the religious nature of the times and the pagan inheritance, this was a massive turf war.

As an example of the approach of the incoming religion we can take the final book of the biblical canon, St John's Book of Revelation, which is full of astrological references.

'And there appeared a great wonder in heaven,
a woman clothed with the sun, and the moon under her feet,
and upon her head a crown of twelve stars.'
– Revelation 12:1

According to Jacques M. Chevalier,[92] this aspect of the Book of Revelation is an appropriation of the dominant astrological thinking of the times with the objective of rendering it subservient to the worship of Jesus Christ.[93] Also absorbed by Christianity were the dates of pagan festivals across the empire, which the incoming Julian calendar took over and made Christian. One unfortunate consequence of this particular appropriation was that ritual days became set on paper by distance and fixed rather than set by looking at the sky and reading the seasons.

Perhaps the most important struggle of the first three centuries

92 Jacques M. Chevalier: op. cit. (p. 34 note 49.)

93 They needn't have worried. True astrological thought is never a religion or a rival to a religion; it is a support structure and one form of initiation into the kind of mind that is a precondition of and receptor to the esoteric aspect of all religions.

AD concerned questions about *what Christianity was required to be*. This struggle took place at the interface of formal Gnosticism and the emerging church hierarchy. René Guénon asserts[94] that Christianity was finally stepped down (made accessible) deliberately so it could become a religion for the masses because all the existing religions of the time had lost their way and become corrupted. It was hoped by the pragmatists that a pure structure supporting an untainted religion may attract great numbers of the disenchanted and ultimately set them back on the Path whereas the initiates only saw in religious structures a diminution of the message, a potential loss of the mystical level of the teachings themselves and of the difficult path into the heart of mystery.

Whatever the story is, it seems everyone had an opinion about the methods of God.

Gnostic Ripples

In 1945-46, writings known now as the Gnostic Gospels were discovered at Nag Hammadi and the library at Qumran was also uncovered. The scrolls are from the early AD period under consideration and are a record of the 'other Christianity' – the Gnostics. This is the movement that lost the struggle with the church,[95] the movement that contained the feminine and an emphasis on inner revelation.

The Dead Sea scrolls from Nag Hammadi so threatened the 20th century religious establishment that even access to them by scholars was suppressed by the church for thirty years – just as the Deuteronomists had suppressed Wisdom/Eve and the early bishops condemned the Gnostics.[96] Why? Clearly, it has to do with the same old challenges to an entrenched power base. As such, it will surprise none of you to

---

94 René Guénon: op. cit. (p. 55, note 79)

95 The church 'winning' is thought by some traditional thinkers to have been part of the plan to step down Christianity.

96 After the birth of Islam in 639 AD, a similar struggle under different names took place, with Sophia recast as Fatima.

learn that the Gnostic Gospels tell us that the early Gnostic 'church' had no hierarchy and both men and women were received as authorities according to their level of spiritual insight. Women presbuterai or elders (the word later contracted to priests) held great influence courtesy of this movement, up until the fourth century.

According to the reclaimed Gospel of Mary, 'the visionary perceives *through the mind*' (emphasis added). For the Gnostic, knowledge – living knowledge – was much more important than (blind) faith.[97]

However, it would be a mistake to set up the formal Gnostic movement as the 'true Christianity', as they developed some questionable mythologies at the same time as they inherited an ancient line of thought. Remember, there was a great deal of confusion following the execution of Jesus. One of the Gnostic belief held that this world is a nightmare created by a second god and is to be rejected. [98] The second god, or Demiurge, was thought to have stolen and distorted the divine qualities in order to create this world, an idea clearly constructed from ruminations on the Fallen Angels. As such, formal Gnosticism saw as its duty a detaching of humanity from identification with this fallen place.[99]

Formal Gnosticism is not the same as the term gnosis, which describes the perception of the inner aspect of any tradition. Small 'g' gnosis would perceive, for example that we do not need to fall into dualism and reject Mother Earth and the flesh; gnosis would perceive that the struggle is

---

97   This emphasis brings Gnosticism in line with the Fourth Way School: G.I. Gurdjieff taught that faith is a form of unconsciousness of 'sleep'.

98   This contrasts with the Hermetic view that we incarnate on earth in order to perfect ourselves and that the world itself is moving toward perfection. This is summed up in the saying: Hermes perfects, Gnostic rejects. See also Elaine Pagels; The Gnostic Gospels, Vintage Books edition, 1981. This popular book is recommended for those seeking to understand the conflict of world views between the early church Fathers and those who sought to recover Wisdom.

99   It could be said that Gnostic error reinforced transcendent spirituality and dualism in much the same way as right-wing Catholicism did. It is the opinion of this author that both of these lines of thought need a good dose of the 'hard to pin down', unified system here called Astrology.

really with our desire nature, our habit of attraction and above all our attachment – the identifications *within* – and does not require us to set aside the beauties that earth provides. Gnosis reveals that, yes, we might indeed be advised not to be attached to this earth: rather, we are *called to be its custodians*. Custodianship is attachment transformed: it is a refinement. Custodianship is awake, attachment is asleep.

Returning to our story: if one regards the finding of the formal Gnostic gospels through the prism of the mythic struggle between the fallen angels and the redemptive angels, one is bound to conclude that the 1945 timing cannot have been accidental. The holocaust of the Jews, the gypsies and the outsiders is surely the endpoint of a literal reading of the term 'One' into fascism, (which also recruited astrology into its predictive occult practices), while the dropping of the atomic bombs over Japan is the negative endgame of the gift of metallurgy. We needed a revelation of a Christianity which, despite inevitable errors and misread cross-influences, held the feminine as equal and sacred, used Enoch and related vision texts and genuinely struggled to get 'inside' the Christian mystery and her language, without pomp or ceremony.

By about 200 AD all of these texts loved by the Gnostics had been deleted from the Christian canon by the new church Fathers[100] – this was when Enoch vanished– and this meant that nearly all of our feminine imagery for God disappeared; and by the fourth Century AD, the war of ideas between the astro-fatalists and the bishops had moved to Rome, where the newly Christianised empire – in an original move – began to execute the less intelligent members of our astrological fraternity for 'predicting the future'. By that time of course, the bishops had also won the war with the various Christian 'heresies' – the Gnostics, Thomasines, Marcionites and Ebionites – and the heresies passed out of history as a force, most likely becoming a trace in the flow of perennial

---

100   They did retain the powerful Gospel according to St John, written in Ephesus at around 100 AD. This was the time when Gnosticism was thriving within the developing Christianity.

wisdom, resurfacing from time to time as elements informing esoteric schools and the understanding of individual teachers.

However, the Glory does not die just because her stories are suppressed; her Spirit passes on through the oral tradition, always disguised, until a moment arises when a web of enlightened men and women are present at one time and the opportunity for a fresh revelation is presented.

This is the subject of the following chapters.

THE HERMETIC HARMONIC
Chapter Seven

And God-the-Mind, being male and female both, as Light and Life
subsisting, brought forth
another Mind to give things form, who God as he was of Fire and Spirit,
formed Seven Rulers,
who enclose the cosmos that sense perceives. Men call their ruling Fate.

– Corpus Hermetica: Book One, Poimandres,
the Shepherd of Men, No. 8,9.

Knowledge Abroad

How does knowledge remain alive? How *does* the individual remain connected with that which he or she once understood? How do we keep our knowledge vital?

The perennial wisdom or gnosis passes through history in waves and surges, much like a harmonic wave in music passing over the electro-magnetic field, except that our wave expresses itself 'in the flesh', as the common ground of a profound, human-centred teaching. Along the journey, as waves rise to visibility, academies are established and the truth that was perceived is taught. Then we forget why we were once so inspired and the knowledge itself becomes routine and empty and insti-tutionalised; and so the Glory passes on until she finds another time and home, and another chapter of sacred history commences.

After the death of Alexander the Great in 323 BC the world that had given birth to Western astrology became Hellenised. Greek was spoken everywhere, even during the centuries of the Roman Empire.

During the early days of the Empire immigrant astrologers living in Rome had an uneasy time of it. Their philosophy, which made the birth chart of primary importance to an individual – above wealth, above class – made of astrology a great equaliser. This was considered a radical and dangerous idea; so in 139 BC all astrologers were ordered out of Rome; but time passes and due, in part, to the contrary influence of the fascination of the Roman intellectuals with 'all things Chaldean', the elite gradually became reconciled with astrology and, by the time of Augustus and the turning of the millennium and Jesus Christ, astrologers had drifted back to Rome.

The first three centuries AD saw a Greece that was the established centre of civilization and which was a hotbed of religious and philosophical belief, debate and crossover influences; in fact, religious pluralism at that time was more extensive than that of Western 21st century religious cults, and a source of great confusion to anyone attempting to make meaning out of life while pursuing their everyday affairs! A number of these movements impacted on or maintained a harmonious relationship with astrology.

The detailed horoscopes that we use were developed during this era (Seleucid) and these combined with the Tetrabiblos of Claudius Ptolemy to create the foundations for a geocentric, physical and predictive astrological art.

There was the legacy of Stoicism, a movement that had arrived thirty years after Alexander as a gathering of the educated in the public square of Athens. Their founder had been Zeno, an inheritor of the Pythagorean Brotherhood and a rival of Plato.[101] Zeno and the Stoics taught (publicly) a philosophy of strict Fatalism based upon the horoscope; however their real contribution to astrology was their understanding of *cosmic sympathy or resonance*, which enables everything on earth – parts

---

101 It could be said that Zeno carried a wisdom line from Pythagoras and the Babylonians. Certainly Plato wanted to get rid of him and to inherit the mantle of Pythagoras himself.

of the body, metals, plants, colours, musical intervals – to be ascribed to a zodiac sign or planet, once the central principle or quality[102] behind the cosmic category is grasped. The Stoics stressed that all of nature is both a manifestation of God and a Unity.

There was the inheritance of Greek mathematics and the Greek quest for the perfect mathematical 'cosmic glue' which holds the universe together. It was this Greek emphasis on a closed, perfect system that robbed those working in the grey area between qualitative number and the human being – as astrologers were and are – of their Uncertainty Principle,[103] which is so vital to prophetic utterance. The belief in perfection kept Greek mathematics in theoretical suspension, for they did not think to test whether or not such perfection actually existed. Their planets orbited the Earth in perfect circles.

Then there was the influence of the Gnostic movement and its clash with the developing church hierarchy, later to become the Greek Orthodox Church; and there was the other kind of priesthood, that arising from the fascination with Chaldean knowledge and its ancient sources. Even the Academy of Plato was still functioning but by then the teaching had become stale: there was thought to be little connection with the original knowledge.

Thus, many influences and schools continued to circulate in the great brew that was Greek thought in the first three centuries AD – and this just as formalism was seeking also to take hold of the Christian Mystery. Within the maelstrom, there were some who recognised the danger: that there had been a loss of the meaning in all their knowledge, and of its vital connectedness, in much of that which was being professed.

It was no doubt in response to this situation that the books of the Corpus Hermeticum were written in the z2nd and 3rd centuries.[104]

---

102 A 20th century exploration of the lost Stoicism and the centrality of Quality can be found in the popular 1970s book 'Zen and the Art of Motorcycle Maintenance – and enquiry into values', by Robert M. Pirsig, Corgi Books 1976.

103 See chapter 15: Neptune the Glory.

104 The British historian Frances A. Yates called the Hermetica 'a refuge for weary Pagans'.

Conversations Along the Road

We do not know who wrote the collection of spiritual teachings and
magical formulas known as the Corpus Hermetica. An educated guess
would be that the books appeared as a result of conversations between the
spiritually educated that took place along the safe, easy roads built under
the Pax Romana, and at Ephesus, which acted as a kind of conference
centre for travellers from all over the Mediterranean. Ephesus boasted
lecture halls, schools of philosophy, temples, synagogues and theatres
and, shortly after 100 AD, had its own library. Recall that The Gospel
According to St John was written in Ephesus around that time. Along
the roads and into the Greek 'centre of civilised thought' would have
travelled some of the great thinkers of the age, no doubt carrying letters
of introduction. They would have sat together in taverns and talked about
the staleness in every branch of philosophy, about the politics of power
and its corrupting entry into every branch of religion.[105] Some would
have become interested in revitalizing and recording the living aspects
of their knowledge before it rotted away under a looming intellectual
decadence.

In other words, there may have been an informal esoteric school,[106]
a web of connections. The Corpus was written because it had to be,
because prophetic thinking and gnostic insight were losing the battle
and the academies could not help.

When the books of the Corpus were gathered together, they were
attributed to Hermes Trismegistus (thrice-great Hermes). Now, there is
plenty of historical material available on the figure of Hermes: that he
appears first in Egypt as Thoth, later in Rome as Mercury and cross-pol-
linates with the functions associated with god figures across the Middle

---

105 Nothing changes.

106 In a sense, all esoteric schools arise informally. They appear to exist for a purpose,
    rather like a magnificent theatre production that comes into life, makes an impact on
    memory and meaning, and departs without a trace. Fourth Way teacher Dr. Phillip
    Groves suggested that the Renaissance occurred because there was such a 'school'.

East such as the Babylonian scribe Nebo – but this is not the point. The Hermes Trismegistus who for at least 1400 years was thought even by church fathers and scholars to have written the Corpus was simply another example of Eastern pseudepigrapha, a name used as a seal of authority and a reference point indicating the nature of the writing– almost a logo.

The living harmonic line known as Hermeticism became such a powerful gnostic vehicle that it eventually provoked the Renaissance in Europe and contributed to the eruption of the scientific revolution.

Poimandres, the Shepherd of Men

It is difficult for us to perceive how pervasive and integral astrological thinking had become by this period. We tend to separate it and make it a 'subject', then we study it in a historical and political context alone. However, astrology was one aspect of a total state of mind; and as the worried carriers of esoteric thought began to record the small books that became the first teaching of the Corpus Hermetica astrological thinking was *an automatic part of their means of expression* along with and comparable to other underlying systems that were a part of the line of glory: Enoch, the wisdom-based Genesis, Pythagoreanism. One is reluctant to call such subjects 'influences' because such a term dries out the lovely living quality of the teaching and the sense one has that it was written down by those who had recovered – in part by a study of these 'influences' – the original perception that qualifies a person to make their own contribution to the body of wisdom literature.

The first book of the Hermetica is called the Poimandres, which means man-shepherd.[107] It is a story of origins and a description of cre-

---

107 This reference does not mean that the Poimandres is a form of Christianity with its Good Shepherd. In fact, the Hermetica quotes no specific biblical stories (or Greek mythological figures). This historical 'Hermes' is also found in the ancient Homeric hymns, where he tends the sheep of a mortal shepherd in idyllic Arcadia. There he meets and mates with a mortal woman. The goat-footed god Pan is their offspring. Pan

ation in the form of question and answer, which becomes a vision and a repeated litany of Praise. Already we are in the presence of Enochian transport, a Gnostic mother-father creator and the Stoic embrace of nature by God.

> *'And Nature took the object of her love and wound herself completely around him, and they were intermingled, for they were lovers.'*
> – *Poimandres, 14.*

We are also in the presence of the seven planetary spheres, ruled by archons, whose 'role here is an oddly ambivalent one, powers of Harmony who are nonetheless the sources of humanity's tendencies to evil.'[108] It seems that the Fallen Angels are also a part of the brew; and the very 'ambivalence' of their function is indeed an indicator that the Poimandres may be an authentic doorway for contemplation, for truth never uses dogma as its doorway nor permits any teaching to become morally sealed off. *Uncertainties invite our participation.*

While a close reading of the Poimandres presents many treasures to the inner life, other books of the Hermeticum take a more occult line and include (for example) magical formulas for drawing down the energies of the stars and bringing statues to life (again). Here the occult appears to be a dynamic accomplice to the work of resuscitation:[109] it is an energy and it is oft-sought by outsiders or heretics.[110] Of course, wherever insight into the structure of power is acquired, there will be a

---

(along with bunyips in Australia and coyotes in America) advocates for nature when we turn away from her and arbitrates and punishes those who misuse her resources.

108 John Michael Greer, introduction to Poimandres, the Shepherd of Men. http://hermetic.com/texts/hermetica/hermes1.html, accessed 17/1/2013

109 I will have more to say about this in the chapter on Pluto.

110 The best known surviving story stamped by the 'hermetically sealed' mind of the times is *The Golden Ass* by Apuleius of Madaura, an educated man seeking release from the mechanical teaching of the times by engaging with daemons and the occult. Many versions of his book are still in print.

tendency to manipulate and control what is thereby found – not always a bad thing and the basis for all medical science – and here it seems that occult magic was a practical offshoot of the astrological doctrine of cosmic sympathy (as it was later in the Renaissance with different results). In the Corpus Hermeticum, the doctrine was beautifully expressed in the well-known 'As Above, So Below' of the Emerald Tablet for, despite some of the more eccentric applications of their belief, the Hermeticists knew in their hearts that the human being could be saved by a visionary knowledge delivered by *the heady cosmic sympathy, which carries the love between God and Nature.*

This is the point of the Corpus Hermeticum.

Tributaries

And so the wave that was hermetic thinking entered into sacred history and, as the official church and its hierarchy erupted into sight, dipped and passed as tributaries into the great underground stream along with the Book of Enoch and many visionary texts. By the 5th century, Gnostic thinking was presented to official history as heresy at best, demonic occultism at worst.[111] Christianity was accepted as a mainstream religion and, in 529 AD the Academy of Plato at Athens closed and its teachers and students moved over into Jundi-Shapur in Persia. The library at Alexandria was destroyed: remnant books were relocated to the Christians at Edessa (the descendants of those who had avoided our Sabians on their lofty heap).

During these times, the persistence required to keep the gnostic eye propped open was often supported by magical practice. 'Dead' knowledge and occult practices are always uneasy partners in the cycle of diminishing returns. In this instance, magic became legitimised for a time by the emergence of a category called theurgy, which is *magic with*

---

111   One problem that the early Gnostic movement in Christianity faced for centuries was that most known records of it were written by the bishops.

a religious purpose and ritual intended to support initiation (whatever that was thought to be). Theurgy was well established and growing by the 4th century – the time when women Gnostics lost their authority at last– and by this means, a polluted form of sacred history was able to move on, much like a disease having to run its course. By the 7th century the Hermeticum and related magical rituals had wound their secret sinuous course back across the border into the east, where it will surprise no-one to learn that they flowed into the well-established Sabian practice, although who reinforced whom cannot be known as both streams professed planetary governance, cosmic sympathy and the animating of statues. Certainly the Sabian library contained books by 'Hermes' (although false attribution to real authors as well as to mythical ones was common at that time). One 9th century heretic[112] from 'Sabian' Islam attributed the entire Sabian star doctrine to 'Hermes'![113]

The tributaries had become a whirlpool at the interface of the Eastern and Western worlds.

Meanwhile, in Europe the Roman Empire had transitioned to the Holy Roman Church: the European world had come under the spell of an exoteric Christianity and its bishops– and here we must pass at last firmly onto European soil.

---

112 The fact that he was executed for 'heresy' inclines this author to take his comments seriously.

113 One reason Hermes had been accepted by some Muslims is that he was thought to have been a 'scientist'.

THE MAGICAL FORGE
Chapter Eight

*'Astrology is (also) an 'art mathematical' and one which particularly
shows forth the glory of God, who made the heavens in His wisdom.'*

– John Dee, Preface to Euclid, 1570

Renewed Lines of Sacred History

In the last chapter, we recorded the entrance of the Hermetic line into
sacred history. During the Byzantine Era that followed (from 330 AD)
the fifteen tractates that make up the Hermeticum were collected into
one document and copies were stored in the monasteries of the East.

The Corpus Hermeticum could be said to be but one line among
those lines of sacred history that lightly touch and renew the human
soul, much as the fine energy of the electromagnetic field embraces the
physical earth. They are like threads in a beautiful woven representa-
tion of the face of Glory or tributaries seeking to flow into the heart of
Mystery.

There are moments, in recorded history and in sacred space, when
a meeting of several of these lines will occur, and if there are enough
prepared individuals at the time, a great flowering of culture will result.
We recognise these periods: they leave a grace for humankind forever.
This chapter is the story of three such lines of sacred history, which
joined forces with the re-emergence of the Hermetic line to enable the
Renaissance in Europe. It is also the story of how astrology came to be
reclaimed – only to be lost again.

We will start with the monks.

Megalithic Monks

It has been told how, at the commencement of the first millennium, there was a struggle taking place at the interface of formal religion, magic, paganism and Gnosticism and that this struggle was throwing out hybrid notions across the East-West divide. However, there was another, surer parallel development. This was the coming of the scholar monks and the gradual *spread of monastic culture from the East into the West.* One effect of this was that many pagan texts were preserved and quietly protected along with religious commentaries, history and astronomy, all within enclosed libraries to be accessed by this new class of Christian researchers right across the Christianised Western world. Particularly, knowledge of astronomy was sought by the monks because an accurate reading of the night sky would correctly set the times for prayer and an accurate calendar would firm up dates of feasts and Christian holy days.

It was during the first 700 years of the Christian era that there was an ongoing dispute over the setting of the correct date for Easter, which is the most important date in the Christian calendar, and in this dispute the advice of the most learned of the monks was sought. The problems were several: there was the story – Easter was and remains the story of the fulfilment of the Old Testament prophecies and the renewal of the world and this intensified the struggle over the date; there was the problem of achieving international agreement as Easter had to be celebrated on the same date everywhere to hold its power; and there was the complex astronomy which needed to bring together the date of the Spring Equinox, the first full moon following this date and the appropriate Sunday; and all of this had to be anticipated in tables as the season of the Lenten fast in the Christian calendar preceded Easter. What a mess.

For a time, it seemed to the bishops in Rome that the most accurate cycle to be used for the tables was one of 84 years because the date of Easter could be recorded accurately over this cycle with an error of a mere 1.28 days, which would necessitate only a small correction, much like that of a leap year. The 84 years was made up of three Metonic cycles of 19 years each plus an additional 8 years. A Metonic cycle is the time taken before the sun and moon return to exactly the same relationship to each other. Those of you who have been paying attention will observe also that 8 years constitutes one cycle of Venus.

Now, the Irish – along with some Britons, particularly the Picts – were quite attached to the 84 year calendar cycle, as initially agreed upon. A probable reason for their enthusiasm was that their monasteries were positioned right on top of the ancient megalithic sites. In fact, there was a monastery at Brú na Bóine which by the sixth century boasted some 3,000 students. These monastic folk were the distant on-site inheritors of our mytho-astronomical mindset. After the great emigrations to the Middle East, those who remained had kept intact their mathematical-storytelling language and this had passed orally to the bards and to the Druid priests. As a result, the Irish did not take to the notion of the pope-bishops-laity hierarchy per Roman Catholicism; their Celtic Christianity was seated in the megalithic monasteries, whose abbots always had grass roots blood ties to the local communities. The Celts even rejected the Roman edict that Jesus was 'God' and instead saw him as a powerful teacher, utterly compatible with the Druidic line of wisdom, and its oft-time expression in bardic parables.

So the stubborn Irish and the Picts did not respond when the Roman Catholics changed their computing of Easter and dropped the 84 year cycle. No doubt the ley lines of time, backing into the mists of the wisdom of astronomer-priests had caused them to recognise the symbolic value of a triad of cycles of the marriage of the sun and the moon capped off by the rebirth symbol of the cycle of Venus. Remember, the monasteries had been placed precisely at the places of megalithic inheritance and Brú na Bóine was a site of a Chamber of Venus. (This is always the trouble with attempted appropriations: often the site ends up appropriating the appropriator.) Of course, the insecure Roman Catholic Church, inheritors of a crumbling Roman Empire, would never acknowledge such a legacy: they only saw an Irish 'problem'.

It was around 590 AD or so, that one 'problem' Irish priest named Columbanus immigrated into Europe along with twelve of his religious compatriots and there he eventually established three Irish monasteries. This Columbanus had been trained by an aging bard and had spent his early Christian years at Brú na Béine, where he had studied the Psalms in depth. In Europe, his monasteries held fast to the 84 year Easter cycle and, in one, Columbanus instigated a service of Perpetual Praise. Columbanus had little or nothing to do with the bishops.

This is how a particular line of sacred history passed into Europe once more, in a loop from the far West onto the European mainland, carrying its archaic astrological symbolism, its scholarship and its praise. It is outside the scope of this book to follow the ongoing astronomical war and its final resolution. Instead, we will pick up the main line of the monastic movement and its hidden libraries once more, after it has passed into the Medieval period.

The Monks Re-open the Books

By the time of passing into the Middle Ages, magic had been banned finally by the Churches in Europe as dangerous pagan thinking and (while there is no doubt that it would still have been practised in webby little nooks and crannies[114]) one positive consequence of the ban was that, for prominent philosophers and scholars at least, there was now room to re-open a discussion about astrology without the fear of being drowned in accusations of paganism. Astrology could be discussed in the context of astronomy, and paganism – officially – did not exist. Now, just at this time, when astrology was being released from its pagan-astrological shackle[115], the Christian churches of Europe themselves

---

114 By mid-13th century, the manual of astral magic known as the Picatrix was in wide circulation. It had arrived in Europe via a Latin translation from the Arabic.

115 The line that released astrology probably started at Charlemagne's (9th century) court fertilized by incoming mathematical material from the East and the desire of the court to reform the stars.

reached a hiatus in development, when it was realised that revitalization was needed if Christianity was to recover its original mission. These churches were now all-powerful as arbiters of 'truth' but these were the feudal times and feudalism was an oppressive, brutal and class-based social system, which had only encouraged the lifeless side of church politics and its corresponding structure.

It was in this context – the desire for reform in the period following the banning of magic – that fresh, radical monastic movements arose because the scholar-monks themselves were struggling to loosen the church's death grip on authority and the monks had available to them libraries of core knowledge including the macrocosm-microcosm world view with its astral cosmic sympathy, Hermeticism, Neo-Platonism and the commentaries upon them by the early church fathers.[116] Now, and within the pressure cooker of mounting church territorial rivalries and inevitable resistance to reform, the monks re-opened their astrology texts in safety and began to construct astrology charts, seeking to predict the outcome of theological disputes and even the date of the Apocalypse! The formal church hierarchy counter-moved by appropriating astrology and describing it as a purely metaphorical language compatible with scripture – the 'your eyes are like stars' approach.

It is hilarious to imagine the monks drawing up horary charts[117] in their cells! Perhaps they met in the cell of Brother Peter as 'The Monastic Astrological Society'. Of course, they had inherited prediction from the Greeks and divination from Babylon and, as we have been at pains to point out, predicting the future was never the point of this prophetic gift. Some key figures of Medieval Christianity did recognise this. Thomas Aquinas (1225-1274) wrote:

---

116 These records (if not the understanding associated with them) were of course known to early official Christendom; even the famous church father Augustinian had believed that Hermes Trismegistus had been a real person whose writings warranted a reply.

117 A horary chart is one drawn up for the moment a question is asked. The theory is that the present moment contains the answer to the question arising within it.

'... very many men follow their passions, which are motions of the sensitive appetite, alongside which passions the heavenly bodies can work; few men are wise enough to resist passions of this kind. And therefore astrologers, as in many things, can make true predictions, and this especially in general; not, however, in particular, for nothing stops any man from resisting his passions by his free will. Therefore the astrologers themselves say that 'the wise man is master of the stars' (*sapiens dominator astris*), inasmuch as he is master of his passions.'

– *Thomas Aquinas: Summa Theologica*

Amen to that.[118]

In sum, by re-establishing the monasteries – peppered with an Irish Druid inheritance – and by re-opening the debate about the place of astrology within a Christian discourse, the monks reinforced a line of sacred history that would prove necessary to support the recovery of the core Hermetic texts in the 15th century. The line even travelled openly for a time: by the early 14th century bishops and popes had acquired astrological counsellors!

Yet there is something else, something far more important here and strongly sensed. The philosophy of cosmic sympathy, the web of connecting principles in a vast continuum, which lies at the back of astrology, may have contributed in a mighty way to the recovery of connectedness in the thought and the art of the Middle Ages. The Glory was able to touch base with the thinking, creating human being once again, in part enabled by a released astrological mindset. The result was a flowering of *symbolic and allegorical art* and, in philosophy, a re-opening up of *the sacred relationship between the eternal and nature*.

---

118 What Aquinas omitted is that, by proper use of one's map and by utilizing the transits to purify it, one can be *helped* to grow beyond the grasp of Fate. We use the astrology to acquire the greater 'free will' of Aquinas.

The lid that had been shut so tightly over esoteric languages was lifting: there was something bubbling all over Europe.

The Jewish Heresy

At the same time as the Christian scholar monks re-opened the pagan debate right in the face of feudalism, in Spain the Jews too were plotting their own brand of heresy.[119]

As the inheritors of the Old Testament laws of Moses, the Jews were aware of their responsibility to maintain God's fundamental instruction to all the Peoples of the Book. However, in addition to this responsibility, some Jews had inherited the knowledge that the Law carried a second, hidden meaning, something found in the Glory as God departed. There had been a second revelation to Moses on Sinai, one that could only be understood and transmitted by teachers who possessed the Angel Body. This teaching re-surfaced in Spain and became known as the Kabbalah. The word means 'tradition'.

Kabbalistic practice involved a search for the mystical meaning of the scriptures based upon the Hebrew alphabet, which was manipulated in order to reveal other words within words, the higher meanings disguised within biblical story-words.[120] Using the alphabet, the Jews of Spain developed a numerical system of letters and a corresponding mystical system of emanations from God, known as the Sephiroth. These combined – the hidden Hebrew, the scriptures, the number correspondences and the emanations – into a kaleidoscopic system for contemplating Divine Purpose.[121]

---

119 Both Spain and southern France inherited gnostic lines of thought: Spain from the Enochian Jews and France from the Pythagorean diaspora.

120 Of course, hidden messages cannot be disclosed by moving letters around; it is required that they initiate him or herself to develop an extended perception across the range of meaning.

121 Like astrology, the Kabbalah was all but abandoned during the Age of Enlightenment, even by Jewish philosophers. 20th century scholar Gershom G. Scholem has written

Then, in 1492, Catholic Spain expelled its pocket of Jews. The Kabbalah travelled as an exiled knowledge to France, Germany, to the England of Henry VIII and as far as Turkey, looking for a home.

They found their home in Italy – where the ground had been prepared.

A Note On Pythagoras

To discover our third and unexpected line of sacred history that impacted Italy, we need to return to the vital 6th century BC once more and to the teacher and mystic Pythagoras, who is a key figure in the dissemination of the perennial wisdom into Renaissance Italy.

Pythagoras was born in the year of the captivity of the Jews to Babylon. As an adult, he was searching consciously for the stream of esoteric learning, and we can surmise that he learned from the Jews in captivity and from the Babylonian astrologers during his many travels. He also lived and studied in Egypt for twenty years and some have put him in India, as his later teaching has many tight parallels with Indian philosophy (although he may have come to this via Persia). Certainly he came home with a teaching on the inner and outer meaning of the four elements and the planets, and his mystical speculations are in harmony with the Hindu Upanishads. His teaching on musical intervals has resonance with the Druid priesthood: we will return to this geographical anomaly in a moment.

It is claimed that Pythagoras had been authorised to begin a school, which would reincarnate the stream of living knowledge as public learning as a blessing for the students of the Western world; and so he came to live and teach on the Island of Samos, which is off the coast of Anatolia.

Yes.

---

that this kind of scholarly abandonment of Kabbalah created a vacuum in which 'all manner of charlatans and dreamers came and treated it as their own property'. See Scholem, 'Major Trends in Jewish Mysticism', Schocken Books Inc., 1961.

Pythagoras and his school taught that light and sound are identical as *qualities*, that both are harmonics in a rainbow of vibration where invisible light translates as audible sound. The Pythagoreans taught that sound is light itself but at a lower octave, a lower frequency. Sound is the stage in the organization of vibrations when such vibrations begin to take form.[122] It is claimed that Pythagoras himself heard the sound of the hissing of the stars when he was in a Shamanic state, and he believed that the transmigration of souls finally ended when each was taken up in ecstasy and transmuted into a star. Perhaps Pythagoras heard the Praise of the stars.[123]

Therefore, music was critical to the Pythagoreans as a manifestation of pure light in the world.

The cithara was the stringed instrument used by the Pythagoreans to work out their lasting musical theories.

Now, at that time Anatolia and Samos had many temples with caves beneath them, which were dedicated to the god Apollo,[124] and Apollo – Apollon in fact – can be traced back at least to 1450 BC, when the Lyre of Apollo appeared on the Island of Crete. This event became the source of the musical stream of the wisdom later taught by Pythagoras and his school.

This is where it gets interesting.

It was across this former period – around 1450 BC – that Stonehenge was abandoned, the final site to be left in the great fifteen hundred year exit of traditional astro-mythological knowledge and its priesthood. The folk of Stonehenge came to Crete. Their systems came with them and were passed orally until they reached Pythagoras and became 'the music of the spheres', translated by his school into the musical intervals that tuned the cithara.

---

122 From a biblical point of view, it would follow that the Word is the first sounding of Light.

123 Later a candidate for the Pythagorean Brotherhood had to reproduce this hissing and to say (presumably in ecstasy) 'I am a star' in order to pass into the circle. (Isn't that wonderful?)

124 More will be said about the purpose of these caves in the chapter on Pluto.

There is more. At the same time that Pythagoras was teaching, the cithara was being played by the priesthood in England and was a royal possession of the Druid line in Ireland – those who had inherited the Irish megalithic sites. So important were the tuning intervals that changing the tuning of the cithara was punishable by death. The major scale upon which the cithara was based became the most important scale in the West and the cithara (and later the guitar) became the descendants of the Lyre of Apollo.

The Lyre of Apollo

What then do we know about this APOLLON, who appeared on Crete with his lyre? What is present in the term APOLLON at this seminal moment behind Western culture? In a remarkable piece of scholarship, Anne McAulay[125] tells us about our early alphabets. These were uniquely constructed in such a way that *letters were also used as numbers and as geometric forms*. McAulay then translates the term 'Apollon' into its geometric and numerical equivalence and she discovers that APOLLON is none other than a layout map of Stonehenge itself! The God, the wisdom of the astronomer-priests and the music of the spheres arrived together on Cyprus as one language. That language travelled. It ended up in Italy, where it becomes our creative birthright.

Ah yes, Italy.

Looking back again from the Renaissance, we find that on Samos and in Anatolia in the time of Pythagoras, pressures and dangers began to arise, posed by an incoming war with Persia. The community of Pythagoras emigrated from their Apollonian temple sites. They found their way (eventually and after some mishaps) into southern France and to Italy – to Velia, to be precise – where they set down permanent

---

125 See 'Apollo: the Pythagorean Definition of God' in 'Homage to Pythagoras', Lindisfarne Association, 1982. I am indebted to this marvellous analysis for such enticing information.

roots in the Western world and established two of our core schools of philosophy.

Thus the music of the spheres entered Italy and there it waited for the future, for the time when the renewed search for their own numerical alphabetical language underneath the Old Testament by the Jews would collide with the reconnected, symbolic philosophy of the monks, peppered by astrology. Such streams are outside of time: they wait, they are seeds.

There is one more astrological tale to tell before we arrive at our Renaissance.

Astrological Stages

If you cast your mind back, you will recall the initial struggle that the Roman authorities had in accepting astrologers and that these had been kicked out of Rome for a time. They were finally allowed back because the intelligentsia (under the influence of the Greeks) became enamoured of the ancient philosophy of the Chaldeans and presumably wanted the astrological inheritors of such wisdom in the neighbourhood.

Now, one of the most prominent of this interested intellectual ruling class was the great Italian architect Vitruvius, who lived in Rome between the late first century BC and the first century AD during the time of Emperor Caesar Augustus, who was his patron. Vitruvius's voluminous book *De Architectura* is the only surviving treatise on architecture from that time and place. Like so much writing of that period, *De Architectura* vanished from history for many centuries, reappearing again in its original Latin in various scriptoriums (notably the great one overseen by Charlemagne); and by the later Middle Ages it was being copied, distributed and re-filed in abbeys by our scholar monks. Later, in 1414, *De Architectura* was discovered in a Swiss abbey by a humanist scholar from Florence, who published it openly and circulated it amongst his Italian associates.

The result was a public revival of Vitruvian architecture in Rome in a movement that was to become known as Neo-Classicism. Numerous architectural works followed and the great Renaissance building program commenced.

Vitruvius was an exceptional artist and spiritual thinker. *De Architectura* contains a very long chapter on astrology that takes in all the basics: the zodiac signs, the planets, the geometric relationships. Vitruvius considered this knowledge *fundamental* to the education of the architect along with an absolute knowledge of the proportions of the human body[126] and, particularly, *his designs for theatre were based upon zodiac configurations.*

Therefore, when Vitruvius was rediscovered, Renaissance architecture came under the requirements of zodiac-based geometry.

This story of the re-emergence of the astrological stage in the European Renaissance containsa compelling footnote.

The High Renaissance Vitruvian architecture and Kabbalah of course found their way to Elizabethan England but later, in the late 16th century. Here, in pragmatic Protestant England, there was available one clear avenue down which the acceptance of both the practical *and* the magical side of number could travel and that was inside the new, exclusive actors' theatres[127] just starting to be constructed, and in their stage productions.[128] Theatre always deals in magical effects and actors and playwrights are open and imaginative folk. So the new knowledge came straight to the theatres and the builders set their foundations upon the designs of sacred geometry. In 1599, guild member James Burbage built The Globe, a theatre constructed from the zodiac. It had not only

---

126 It was Vitruvius, not Leonardo da Vinci, who first illustrated a man in a circle with his arms and legs extended to illustrate perfect proportion. Da Vinci, who was in the Italian circle of admirers of Vitruvius at that time, re-produced the diagram.

127 Prior to this time, actors had shared buildings with bear baiters, wrestlers and circuses.

128 The Magus John Dee, a favourite of Elizabeth the 1st, built flying devices for the theatre and his follower, Robert Fludd (who was also a hermetic doctor or Paracelsian) became an expert on acoustics for the theatre. Vitruvius was popularized in England as a result of writing by John Dee.

foundations in the form of interlocking grand trines but the detachable ceiling above the stage was painted with planets and the signs of the zodiac. Within this framework of 'As Above, So Below', the British actors spoke the inspired words of Shakespeare, and in rehearsal they associated with the purveyors of High Renaissance reform. This was an extraordinary blend.

Therefore, the astrological stage emerged in London as the setting for humanity telling of its struggle with the cosmic order and Fate, spoken in the powerful words of Shakespeare.[129]

This was an authentic ritual, beyond time. It is with us still.

Let us return to Italy.

Hermetica Ahoy!

One cannot have a major flowering of beauty and a chance to bring the world alive again without we have three factors. In the first instance, there must be a stable setting based upon tradition; then we need an impacting wave of vitality and, finally, we need an experimental ferment to enable the vitality to penetrate the space. During the Renaissance, it was classical, zodiac-based architecture and the music of the spheres that provided the correct setting; it was the Kabbalist ritual that brought in the ferment, and it was the re-surfacing waves of the Corpus Hermeticum that constituted the vital possibilities.

How did the Corpus arrive in Italy?

Now, there was a translator in these times (15th century) and that translator was also a bishop of the Catholic Church in Italy, and he was deeply disgruntled with the rotten state of the church (the old question, 'how does knowledge remain alive?' and so on). This translator was Marsilio Ficino, and he was one among many disillusioned religious

---

129 For a wonderful story about what happened when the British actors went on tour to Bohemia, with their zodiac and a performance of 'The Tempest', see Frances A. Yates, 'The Rosicrucian Enlightenment', Paladin 1975.

thinkers of the 15th century, thinkers who were *looking for something*. The old academies had become vehicles for Aristotelian philosophy taught by rote, while a rising humanist intellectual class tossed studies such as astrology into the streets, as mere fortune telling. This mattered to Marsilio Ficino because *he was also an astrologer* and he possessed the astrologer's mind.

Now, in the 15th century Cosimo Medici (a member of the legendary and all-powerful Medici family) was in the habit of employing monks as travellers whose mission it was to search out ancient texts through their connections in the powerful network of Christian monasteries and their libraries. In 1460, one such monk returned from Macedonia carrying the works of Plato and the Corpus Hermeticum and, in a stroke of destiny, Ficino was asked by Cosimo to put the works of Plato aside and to translate in the first instance the Corpus Hermeticum; and when the new manuscripts arrived (and the Divine Wisdom whispered to him that the time had come and all the lines of sacred history were in place) Ficino read the manuscripts and underwent a revolution, a reversal in his thinking.

The Corpus Hermeticum was translated and entered the Kabbalist, Pythagorean and Astrological ground of Italy!

News of the translations reverberated through the circles of the hungry elite of Italy and later far beyond. From his position as bishop and based upon the seemingly overwhelming evidence of a superior pagan wisdom, Ficino began to argue for a revolution in church ritual and philosophy and for *the re-integration of pagan thought as a means of saving Catholicism*. Recall that these were times of the looming schism between an emerging Protestantism in northern Europe, which would be violently opposed by the Catholic south in moves that would lead to wars, purges and witch hunts. Martin Luther nailed his challenge to the church door in 1517; Copernicus finally announced to Europe that the Earth revolved around the sun and so undermined the entire church rationale in 1514 (not published until 1543). Henry VIII divorced Catherine and created the Protestant Church of England in 1533. It was an incredible age.

Meantime we have in the esoteric intellectual circles of Italy an *excitement*: a pure geometrical and astrological architecture, the recovery of a corresponding Hermetica with its cosmic sympathy – and all nicely spiced by the drift of Kabbalah from Spain with its compatible secret number system and emanations theory.

No wonder there was an explosion of ideas and beauty.

Unintended Consequences

However, when any form of 'liberating' knowledge is discovered by a bunch of disgruntled, frustrated men, and when that knowledge implies the possibility of the ritual manipulation of divine forces to achieve a set result – well, of course, who can resist this kind of power and ultimate control? Who in his secret self would not say 'churches be damned! *We* know more about divine power than you'? Yet those who were excited by the new Paganism were religious men and committed Christians. It was from this conflict in the mind that the strange elixir that was theurgy – magic with a higher religious purpose – was reborn. Magic came out of the darkness and re-entered the practice of eminent public Christians.

Bishop Marsilio Ficino was more than aware of the problem of the occult. He knew that Saint Augustine had condemned the magical practices of the Hermetica in the fifth century; indeed, he quoted Augustine in public while practicing his theurgy in private! In whatever manner the call to invocation and the daemonic response was dealt with, it was too late. The cat came out of the bag as the manuscripts circulated, and inevitably the old question of spiritual agency was re-opened: do the stars have daemons? Are the stars in control or can their daemons be set to work in the world via ritual? Did Jesus Christ cancel astrology or are his angels really stars? By the beginning of the 16th century, *the figure of the Renaissance magus had been born* – in reality, in the collective imagination of the public and in what would become the fearful minds of Old Power – and as time went by, these Magi travelled across

Europe, forming a powerful web of esoteric knowledge supported by new religious ritual.

The Magi travelled Europe for 150 years – purveyors of astrological thought, secret Kabbalist conjurers of angels, a living ferment striving for a reborn world – and occultists.

A QUESTION OF WITCHES
Chapter Nine

But fate can be also unpleasant or difficult. In this event, however, there are means for isolating oneself from one's fate. The first step towards this consists in getting away from general laws. Just as there is individual accident, so is there general or collective accident. And in the same way as there is individual fate, there is general or collective fate ... All this is connected with liberation from personality.

– P. D. Ouspensky

Poisoned Arrows

We will leave for now our lines of Glory – our Pythagoreans, our Enochians, our Hermeticists and their astrology, our Angel Body teachers – and we will follow instead the line of the occult that came out of the shadows of Renaissance creativity and recoveries. For it was the occult line that served as midwife to the times in which we now live; and paradoxically, *it was the occult that ruined the ground on which astrology might now stand.*

Occult practices will always arise when notions of control or power couple with ego and enter the domain that is rightly graced by the whisperings of Spirit. Any possible understanding of the Glory is polluted by these things. In fact, one could say that when the divine feminine is denied and control of her uncontrollable Presence is sought, there will be a diminution of *both* the masculine and feminine poles in our world.[130]

---

130 Clearly this does not mean men and women as such but refers to harmonious principle.

Put plainly, the masculine will become Ahrimatised, it will place materialism above spirit and become institutionalised and inverted; and the feminine pole will become occult and obsessive. One implies the other.

At the time when our occult Magi were travelling Europe, seeking to re-enliven the Ahrimatised religious establishment, there came about a Plutonian event that struggled and ruptured in the social underbelly, as the result of the inability of these negative religious streams to rise. This was the horror of the witch scare, to which even the most conservative historians attribute the deaths of nine hundred thousand women.[131]

This chapter recalls the history of witches and the astrological meaning of the witch scare.

There have always been individuals in small, rural villages who keepsake a particular knowledge – of local plants for healing, of sympathetic charms to offer one who is seeking a lover, advice about rivals and families. They have been mostly poor people trying to survive and harmless in the context of real political power. Yes, they are superstitious and have crazy 'secrets' and ways of manoeuvring in the village but (up until the late Middle Ages) they and their practices were treated with amused disregard and dismissed as the idiotic legacy of paganism by the officials. On the ground, such folklore signifies the longing in every heart for knowledge, for mystery, for healing and for a territory of power to call one's own, regardless of class and circumstance.

However, in Renaissance Europe something terrible happened to these women (for they were mostly women). They were caught in the crossfire between the religious arrows of the church and the magic-tipped arrows of the reformers, caught in a war of ideas that appears to have had nothing to do with them.

---

131 The figure ranges from nine hundred thousand up to nine million (including those who died in prison), depending upon the vehemence of the historian's feminist argument. Whatever the measure, it was a holocaust against women.

Jumping at Shadows

According to English historian Hugh Trevor-Roper, the witch mania was 'a history of collective cruelty and credulity instituted, inflamed and prolonged (though not always controlled) by organised religion';[132] it commenced in the late Middle Ages in Europe and received its final organisation by the late 15th century. As the reader will realise by now, this was the time of the first rush of recovered Hermeticism (the Corpus Hermeticum had been set before Marsilio Ficino in 1460) and the travelling Magus, and as these Magi began to form a tight web made up of exchanges in knowledge, practices and perception, those with formal powers within the church hierarchy came under siege. If they were not in fact under siege, they thought that they were which amounts to the same thing when one is considering a response. Further, some of the Magi were close to monarchs, were reformed churchmen and aristocrats themselves and this too increased the apparent power of their argument. They might really make an impact, given who they were!

There were many in the churches bloody-minded enough not even to consider a dialogue about reform, no doubt addicted to preserving their own position at the pinnacle of the hierarchy of ideas. These began to cast their nets wider, hoping for a suitable scapegoat for their unacknowledged paranoia. In reality, they wanted to prosecute the Magi as sorcerers and heretics but they did not dare do this – yet.[133] So they widened their definitions, remaking the superstitions of the defenceless rural poor as a devil's conspiracy. Their charms were recast as witchcraft, a heresy among women that required purification by fire! As Trevor-Roper has recorded, everywhere the same story was told and in the same order: *first* arrived the persecutors and only *then* was the heresy uncovered. Inquisitors 'discovered' witches in Hungary, in Scotland, in

---

132 H.R. Trevor-Roper: 'The Persecution of Witches' in 'Light of the Past, a treasury of Horizon', American Heritage Publishing, p. 163.

133 That came later. The Italian Hermeticist Giordano Bruno was burnt at the stake in 1600. John Dee was ostracized and died in poverty.

Germany, arriving in villages with their instruments of torture.

Of course, Jungians will and have had much to say about this, rightly recognizing the projection of the suppressed shadow of these boys-own-spirituality church men; and feminists will recognise the ongoing dynamics of sex and class-based power. However, looked at astrologically we need to pick up as well on a ridiculous justifying piece of reasoning put forward by the church men. One of their primary theories stated that through their Kabbalist practices, our travelling Magi reformers had released demons and these demons had *passed into the bodies of women*: therefore, the women had to be burnt.

Let us look at this twisted logic from the point of view of astrological thinking.

Saturn, Pluto and the Wombs of Women

Pluto is a fertilizing force. Conception and birth are common under transits of Pluto to the natal chart of one or both parents, even to the charts of a grandparent when he or she holds the first loved grandchild with that bright and familiar look of renewal and delight. Life renewing itself, a living knowledge of our own mortality set within the miracle of collective regeneration through the ages: a mystery.

Now, Saturn has many remarkable virtues but regeneration is not one of them. Saturn is a profound thinker about and within structure and, in seeking the Mysteries, he seeks them firmly within the structures that frame our world, in matter, in the use of time, in the working of the body. Saturn is the true research scientist. Often Saturn-ruled individuals utilise their very lives as a testing ground towards obtaining a concrete objective. They know that only a fool would claim that our world has no structure or that we do not need organization within these laws to realise an ambition. The negatives that sometimes arrive under Saturn transits are mere manifestations of a weak spot that has arisen because one has not paid practical attention to it: we are being told that we need more experience in this area if we are to command the laws of life. Being in control is important to Saturn.[134]

It can be imagined that Saturn, having mastered and participated in the practical structures of its time, is a little reluctant when it comes to letting go its position. Saturn's grip on the laws of life can lead those identified with the planet to believe that they *are* the laws. In fairness, there is a certain justification for this; it must be heartbreaking to put in the patience, the thought, the learning from experience, the responsibility, the exposure to criticism by the 'dreamers' in one's society only to find oneself removed and the fruit of one's labours seemingly stolen by those who have not done the work (on Saturn's terms). Saturn does not really understand that there are other values, that the Glory does not require a conventional structure and that the activities of the Glory cannot be predicted. For, over time, structures inevitably degenerate and the folk who made them and hang on, white-knuckled, end up as walking dead, while those who continue to work within them become bean counters. Ask any public servant.

At such a moment, a trapped Saturn needs a crisis to allow the re-entry of holy Life. Enter Pluto.

Every person knows the feeling of fear, its physical manifestations as perspiration and changes in breathing and heartbeat. In astrological terms, the failure to recognise the possibilities of Pluto is a common enough problem and Saturn is particularly blind to Pluto because Saturn is so attached to his earned (and unearned) status. An unevolved Saturn will not interpret Pluto as an opportunity that requires a parallel death of an outworn aspect of our being; Saturn regards Pluto and breaks out in a cold sweat. He retreats behind the framework of rules and laws he has helped to maintain, where he sharpens the cruel edge of his structural

---

134 Saturn transits become helpful when one is resolved to finish a project properly; in fact, it can be said that there will be little trouble from Saturn if the planet is utilized in this way. Saturn and completion are virtually synonymous. It is only when we have put the final touch upon something that we can see clearly that it provides the foundation for what is to come: we now have a foundation and we step onto it in order to see the entire horizon. If something collapses, the foundation was not properly constructed.

weaponry. However, this is of limited help because Saturn is left with the problem of the cold sweat – how can he get rid of the humiliating smell of his own fear?

Of course, a force like Pluto never just departs because one retreats from it. Pluto has a job to do and that job is death and transformation, leading to renewal. Pluto will keep probing for weaknesses until she finds a point of entry. It may suggest to Saturn that fear itself can be a stepping stone, that cold sweat can be transmuted into a kind of sap that will enrich the transformational process and bring a life and flexibility into the character and body. The key is to enter into one's fear and remain present, to enter the cavern with its echoes and streams.

Transformation of emotion is a task of Pluto.

At this crucial time, there will be some in the tribes of Saturn who *will know* and who will re-make their ambition, so that it is no longer a question of status but of character: they will become ambitious of self-purification and develop a clear aim. This revelation of a fresh form of ambition is *the* doorway for Saturn, the metanoia, the change of mind through which one may enter the Mysteries.

However, there are so often the other, the less evolved, Saturnians present in the transformational story, and these cling to the dead forms over which they appear to have control and from which the Holy Spirit departed long ago. Such were the bloody-minded in the upper echelons of the churches in the 15th century. These shrank from the humiliating smell of their fear, their cold Saturnian sweat. Unable to transmute it and free themselves and unable to bear the pressure of their paranoia, the church fathers transferred their fears and night sweats onto others. The witch scare was a *provocation*, a command that others would feel their fear for them and enact their death.[135]

Why village women? As already remarked, it is politically obvious

---

135 Those with Pluto in opposition to one to the personal planets need to be particularly careful of this tendency to provoke others in their lives into taking on their own necessary crisis and 'death'. In the long run, this can only weaken the character and lead such people to become fearful of their own, literal death.

to choose the weakest and thus the most easily frightened. However, there is another mystery present and that is the mystery of the womb itself. For the womb of the woman does not only conceive the baby that regenerates the race and she does not only measure time in cycles (although she does both of these things). The womb as Receptor is capable also of taking in *collective emotion* – or rather a zone just out from the physical womb has that function. Certain women enact emotion on behalf of others, particularly on behalf of those they are close to; and learning to protect the womb is a preliminary step for those women who step onto the spiritual path.

Thus the church fathers were accurate at a psychic level – even though ignorant and cruel – when they accused women of 'receiving demons' – only they were the demons of their *own* untransformed emotion, sent forth to bring the poor into the same fearful state into which the fathers had brought themselves. Like ships passing through a lock, the equal waters of fear enabled the panic of the fathers to pass into the bodies of the spiritually receptive and politically defenceless.

Nuts.

The first Papal Bull condemning witches and authorizing their execution was sought by the Dominicans of the Alpine valleys, who needed the stamp of a higher authority in order to justify outrages already being perpetrated by them in the name of power. As soon as they received official permission, of course they used it to set in place an hysteria about witches among the populace. The Witch Bull acted also as a sledge hammer to crush their more reasonable opponents in the ongoing territorial war of religious command; thus the virus escaped its potential containment and raged into greater Europe.

Naturally, the intensification of the witch scare served as an energetic emanation of the insanity of its promoters – for they were insane. It was a madness of Saturn under siege, full of weird logic serving cruelty, of superstitious hallucinations about devils and diabolical daughters riding broomsticks against a sunset sky. 'Witches' were accused of drenching themselves in devil's grease in order to slide from their houses through keyholes, cracks and chimneys. They joined diabolical parties where they feasted on exhumed corpses, kissed the anus of the Devil and danced about his Saturnian, goat-footed form or had sex with him as he shape-shifted, a filthy protoplasmic incubus spread spiritual disease...

The merest sticking point was required to glue the accusations: she lived alone, she had a cat, another villager whispered about her, she was old and had a big nose... it would make a wonderful cartoon to educate the young on the extremes of idiocy – if it had not occurred in actuality, in all its terror. Trevor-Roper informs us that whole parishes were de-populated and families exterminated, that in one particular village only one woman was left alive after the inquisitors had paid their visit! On the way, laws were changed to permit the inquisitors to burn women who demonstrably had harmed nobody, labelling them 'good witches' before dunking them and sending them to the stake.[136]

Why did the scare come to an end? There are many ways of approaching this. Perhaps generations passed and the disease ran its course; perhaps the occult itself (and the fear of it) finally accomplished its task and regenerated the social norm sufficiently to let in the birth of science and, a little later, to reveal the light of Uranus. Perhaps on the ground there was the inevitable swinging back of the pendulum to reaction and then sanity, to rational argument and doubt possibly driven by the new humanists and those who claimed the moral fruit of genuine spiritual transformation (always associated with conscience). Then the incoming Protestantism was not granted the power that the Catholic Church had: [137] it seems the people had had enough of this. Perhaps new leaders arrived. Perhaps all of the above.

---

136 Interestingly, one voice defending these women was that of Henry Cornelius Agrippa (1486-1535) who was a magician. He himself ended up receiving the main accusation of 'evil' in his time and, although he was not burned, his reputation was utterly destroyed. Yet in terms of the witch scare, he was a moderate and intelligent. Later, he became the model for Goethe's 'Faust'.

137 Protestant Holland was the first state to end the persecution of witches – even to doubt their existence.

There was, however, one key event that fell like a cliff collapsing into the narcissist pool of both the fathers and the invocating magicians and which removed at last, albeit slowly, the fear of the occult and made the manufacture of witchcraft unnecessary.

A BLIND DATE
Chapter Ten

With calm and certainty, enjoy God in any way and in all things without having to wait for anything or chase after anything. To this end, all works are done… if anything does not help toward this, you should let it go.

– Thomas Merton

In the past there were, and may still be, monasteries where Angel-Training – deliberate assumption of the higher mental, emotional and moving centrum activities or one or another of the angelic presences already working in the Work – was performed.

– E. J. Gold

Blind History

It is common knowledge that the 17th century in Europe marked a massive historical turning point in the West, which saw the birth of science, the demotion of the authority of the Church and the casting out of magical superstition from public credibility; and that this revolution in how we frame our ideas has brought about phenomenal achievements while, at the same time, leading modern Western man into increasing isolation, various forms of excess and a crisis in our relationship with nature. How can both of these outcomes prevail at the same time and in the same (sooty) breath?

Now, the centuries that brought the Christian religious wars, witch burning and Renaissance magical revivals also saw a number of human-ist colleges of learning begin to open across Europe. The humanists, formal scholars who taught and studied at these colleges had little to do with the prevailing religious atmosphere (and who can blame them?) Their concern was, as ever, with an increase in overall knowledge. A good example of college curriculum comes from one of the earliest in Mantua, which boasted the classics, philosophy, history and mathemat-ics, and a little music and dancing. It drew students from the elite of Italy, including two women from the aristocracy. The colleges were not universities as such and so did not teach theology, law or medicine; and astrology was dismissed.[138]

A requirement of the times stated that there had to be a system of scholarly checks to verify all knowledge – a good idea, given what was going on! – and so the humanist scholars of Latin had developed a series of tests with which to try Latin texts, so that they might be placed in historical order to tell the story of former times correctly, along with the history of ideas. The tests, of course, were able to be applied to texts other than those in Latin. The tests were straightforward enough, involving a study of style and idiom and, particularly, of references made within the text such as quotes from previous authors, citations or commentaries on current disputes. Yet strangely, this long-established method for cross-checking claims made within or about a text had never been applied to the 'Corpus Hermeticum' by 'Hermes Trismegistus', which had become the justification for 200 years of magical influence, hence Catholic reaction and witch hunts!

Why had nobody checked the Hermeticum? Why, given the resources and scholarship within the Catholic Church itself, had nobody thought to appoint several of the scholar-monks to trace the authenticity of these writings? It seems that history itself had produced a blind spot and that this blind spot, in turn, had shaped an era.

---

138 Ironically, one of the students in Mantua was Pico della Mirandola who later became a magus and a friend of Ficino's.

First Light

Isaac Casaubon (b. 1559) was a prominent Swiss scholar, specializing in the classics and church history and an expert on verifying so-called historical writing. In the early 17th century, Casaubon went to work on the Corpus Hermeticum, publishing his finding in 1614, the year of his death. He pointed out that this Hermes Trismegistus, if he existed, could not have preceded Moses or even Plato as claimed by the Magi because 'he' was writing in the style of early AD Greek and, further, neither Plato nor Aristotle nor any of the authorities assumed to be writing after his time had ever mentioned him! He pulled apart the plagiarism of the Poimandres, its sideways quotes from St John's Gospel, from Plato, Genesis and the Psalms plus direct quotes from later Greek authors and concluded that Hermes was really some sort of gentile prophet. (Note that Casaubon still believed the Corpus to have been written by an individual male.)

A new clash, one of intellectual frameworks, had appeared.

The developing struggle was not the usual one between opposing patriarchs over who was to control the interpretation of meaning, over 'whose authority'. For, as the knowledge of the non-existence of Hermes Trismegistus began its slow ripple through the circle of men concerned with such matters, it combined with and reinforced other ideas in play, acting as a kind of ferment for new *courage* as well as new thought. Indeed there were other individuals born into these times who truly despaired of the superstitious shadows being pedalled by both Church and Magi and these, too, craved a dialogue about ideas free of such shadows. Kepler covers this time, as well as René Descartes and an interesting materialist thinker (in a recognisably modern sense) Marin Mersenne, who opposed any philosophy that could not be materially proved and who regarded astrology and related subjects as products of insanity. Casaubon's proofs found their way into the hands of many of these new rationalists and became an absolute counterweight to the arguments of the Magi – effectively bringing down Hermeticism and so

Renaissance magic in the short term and, in the longer term, demoting unchallenged religious thinking of all kinds *including astrology*. One could say the baby was thrown out with the bathwater...

Let us ignore the emerging materialism and look to one feature in the astrological background to this transition.

The Rings of Saturn

In the world about which we are speaking, Saturn was thought still to be the outermost planet, ruling over the melancholy temperament[139] in man and limitations within the universe. In those days, it ruled two astrological signs, Capricorn and Aquarius, and was considered to function at its best when positioned in one of these signs. Saturn in Aquarius rules Mind itself, the internal 'program' that enables only man to reflect upon ideas and to find formulas. Saturn in Capricorn rules Time and Space, the factual setting in which we find ourselves with its geology, location and laws. Both govern limitation of one or another sort, for what is limitation if not reality? Thus understood, these ruling features passed from the Ancient world on into the Medieval and Renaissance languages of the esoteric circles.

The rings of Saturn were unknown until Galileo saw them through his telescope in 1610, which was a year in which Saturn was in Aquarius conjunct the Asteroid Pallas[140] and the technical planet Uranus was in the double-sided communicator Gemini square Neptune, giving an archetypal discovery in communications that would also create confusion and division. Chiron, on its journey to connect the transpersonal

---

139 Marsilio Ficino remarked that 'philosophers will come to know their Saturn'. Ficino deliberately published his own translation of Plato when Saturn was in conjunction with the planet ruling knowledge, Jupiter.

140 Saturn was conjunct the asteroid Pallas (which asteroid relates to the perception of patterns) across the summer months in the Northern Hemisphere, when clear skies and mild weather would have encouraged Galileo to plot the stars.

planets with personal understanding,[141] was at 0 degrees Capricorn, an extremely powerful position for any astronomical body.

Galileo also discovered the satellites of Jupiter and – rightly as it turned out – suggested that they could be used to give the longitudinal position of a traveller. The suggestion was made in 1616, the same year that the Inquisition declared as heretical the new solar-centred science and demanded the withdrawal of the publications of Copernicus. Galileo had written to Johannes Kepler, telling him that he believed, along with Kepler, that the Earth revolved around the Sun. It was a time of turning in many senses.

It has been briefly mentioned that, in astrology, the rings of Saturn relate to the nature of thought. Therefore, let us look at Mind in the Saturn in Aquarius sense for a moment. For all its potential, it is probably most true of Mind that it runs on a track of repetition: old ideas and memories fold back upon themselves, unresolved relationships, regrets, wishful thinking circle around until we either freeze into a 'type' or develop a functional loss of one kind or another. One could say that, unless confronted, thought will become a blindness and great humanity a predictable machine reacting, reacting…

The rings of Saturn are believed to be the debris of a moon that has already been broken apart and pulverised. So we have a clue about the end of the circling and overly receptive mind.

In the language of astrological storytelling, the discovery of the rings of Saturn was an invitation to humanity from Mystery, asking us to begin to speculate on the nature of Mind *in its limited sense*. In what way has repetition frozen the mind? Is it time to question fixed formulas and doctrines, which we have been repeating and repeating at both the official (Tenth House, ruled by Saturn) and public-scientific (Eleventh House, ruled by Saturn) domains? Saturn is a master of melancholic pondering. He is not particularly creative in any radical sense; rather, he seeks to unfold patterns that are already present and 'true' and his journey is a

---

141 Chapter 11 will open out this statement.

respectful one. It would not occur to him to add a wild element to his reasoning and so remake it as experiment, to break out of the circle.

Respectful journeying is a pure quality of Saturn *and* one of his blind spots, for we must examine closely what it is that we respect and why.

Thus it transpired that Saturn in Aquarius, with his emblematic blocks of set abstracts or (interpreted another way) his necessary sane confinements of mind and his presentation of protective, patriarchal fixed formulas, was beginning to be asked to consider the rising blood of a new, creative manifestation of thought and to release its creativity from Saturn's automatic pilot.

Uranus was using the gateway of Aquarius, and even as the rings of Saturn sought to handcuff the inevitable, Uranus's methods began to break into view.

The Great Man Fallacy

Those of us who love the doctrine of cosmic sympathy, that which creates the world as an interplay of spiritual forces along lines of the principles blessed by the Glory, are almost disheartened by the public history of the 17th and 18th centuries, when astrology was marginalised and became part of talks given by funny little esoteric societies and the practice of eccentrics or passed into the truly dreadful if harmless columns of the Sunday press and glossy fashion magazines.

Almost disheartened but not quite…

If we trace the means by which the new disrespect for astrology arose, we find that in Hermes Trismegistus the doctrine of cosmic sympathy had become inseparable from the doctrine of the 'great man'; hence, when the man was proved to be not only 'great' but in fact a fake, his doctrine likewise fell. Yet the *content* of the various texts still existed – and exists. The Corpus did indeed quote Genesis, the Psalms and Plato because those who wrote it down in 200-300 AD saw themselves as custodians of the perennial wisdom during complex times and they had inherited the old tendency to write under the name of an 'authority'. Of course, this does not make the Hermetica some sort of alternative Bible: it makes it one legitimate contribution to the gnostic canon, with all its light and its limitations.

The private wish to become a Great Man and the search for appropriate models is an unfortunate blind spot of a certain class of men (much as fantasies of glamour infect certain women). In this search, such men never factor in the way in which the adopted great male candidates have, for instance, been nurtured by their mothers as children[142] or how they might have related to others as adults; and as the search for the ultimate Great Man recedes into history, the candidates for the position of Great Man model are more and more misinformed by speculation and then by myth until we arrive at God the Father, the very seat of anthropological projection whence the male projects his Great Man longings and affirms his own argument in the guise of God's blow-back blessing on his candidate-sons.[143]

This does not mean that 'God' does not exist; it means that we need to recover the word free of ego projections and make the space for God to be revealed to us as moments of grace, uncontrollable enchantment and the living, unmarketable beauty of Nature and her humanity – us.

Our group of patriarchs had projected their longings for an alternative to church authority back into history and, finding Hermes Trismegistus, had found it convenient not to question his credentials. As always, they found just what they were looking for. Gee whizz.

Another problem that always accompanies the Great Man fallacy is the problem of false hierarchy. Now, hierarchy does exist: any idiot will tell you that love, beauty and wisdom are 'above' (have greater weight,

---

142 The humanist Francis Bacon, who helped set in motion the modern approach to knowledge and who avidly avoided the Magi, was raised by a puritan mother. Kepler, who scrutinized scientific facts trusting that they would further reveal the hand of God, had a mother who later came under accusations of witchcraft. So it goes at the breakfast table of the young genius.

143 Another problem for the West with its Latin-based languages is, of course, the carry-all term 'God' itself.

are more important than) gratification, fame or expertise; and it is certainly true that some human individuals have become more evolved than others within the same social/cultural milieu. The point is that, without the web of relationship and the honouring of the Glory – these being without hierarchy – any *living* hierarchy cannot be disclosed and we will compensate with the lifeless, 'diabolical' version.

Interestingly, Angela Voss[144], an astrologer and academic, has suggested another simple model for God as an alternative to the tired, old hierarchical legacy. Voss points out that the old model goes back to Aristotle, who envisioned creation as a ladder with the Divine Mind at the top, followed by (in descending order): the angels (in strict Pseudo-Dionysian order, much like a public service with wings), the fixed stars, the seven planets, the human realm, the animal, plant and minerals kingdoms. She says that the result of this thinking is to place God well beyond the reach of His creation and gives rise to the dualistic thinking that degenerates the earth.

Amen to that.

Voss herself uses an onion model, where the divine principle (the sun being its symbol) is at the centre of the onion and its energies penetrate the onion rings, layer upon layer. This spherical model creates God as immanent, as permanently within every part of our world.

There can be no doubt that when the revival of astrological architecture, the Kabbalah and the tracts of Trismegistus emerged in Europe, the pre-conditions for a fresh embrace of the Glory and her 'As Above, So Below' connecting web were created. For a time, there would have been so much excitement in the hearts of the reformers. However, the old longing for religiously-justified ladder ascension could not be suppressed in these Very Important Men and so we ended up with ritual formulas for the control of angels versus equally controlling reactionaries.

The true perception, which is here called the Glory, could not be found.

---

144   Angela Voss: God or Daemon? Platonic Astrology in a Christian Cosmos, in the Journal of the Temenos Academy, No. 14, 2011.

Gutting the Glory, Preserving the Magic

For a non-existent great man, Hermes Trismegistus certainly had an enormous impact upon Western history and not only upon the 200 years of magic and burnings. You see, the Magi – unknowingly – trained an intellectual elite who were *able to will themselves into a command of nature using number as the doorway*, even if it was a spiritual number-continuum that they were expressing at the time. In fact, the number-based Kabbalist rituals of the Magi, when taken as an attitude, have all but defined the modern attitude. Theirs was number applied to the inner realm of Nature and backed up by an excited ego. Surely this represents the birth of our science.

Now, Renaissance magic had the *appearance* of Glory – it looked for a time as if the Magi might be encouraging Her to return and bring life back to religion and certainly She may have blessed the work in the beginning – but the Glory is not magic.[145] Magical operations are, in the end, based on a desire for scientific *power*. They are ritual operations intended to enter into the sources of power energies and to manipulate them. The modern age stole what it could use from magic while discarding what it could not, such as invocation and the Hermetic consistency of principles across the three worlds.[146] Above all, science inherited the attitude of the Magus and the training of his imagination, which was applied by Modern Man in various ways including the math of nuclear fusion.

So much for the Hermetic teaching that man is made of Light and Life, and the (renewed) Sabian/Platonic call for man to know himself.

Right at the beginning of this transition to modernity we find, for example, Descartes in 1619 rightly throwing out the prevailing magical dimension in mathematics - and accidentally throwing out the living Glory as well, expressed through the qualities. Descartes is left

---

145 One function of Neptune is to make one thing look like another. In this case, magic looks like the Glory and so both become rejected at the same time.

146 Heaven, Earth and Man.

then with his mathematics but realises that the shifty mind cannot be expressed through his numbers; and he does not know what to do with God, yet he is a religious man. Therefore, he separates mind from matter, which can be rendered mathematically, and he tosses God out of the universe as well and gives him the boring watchmaker role where He rules over a mathematical/mechanical world. Step by step, God is again distanced from reality and has ended up at the top of a refreshed ladder.

Really, it is so hard to get these men to give up their ladder!

Curry asserts that modern science has borrowed heavily from the ideology of magic, elevating another kind of male Magus who is using number techniques to enter into and manipulate Nature and that Man's discoveries are permitting others to apply science to Mother Earth without respect or consultation.[147] Curry's argument is not against the useful application of discovery; it is against pure usery without the accompanying sensibility and consultation with Nature that the Glory bestows.

We took out the Spirit and Her astrological wheel – but we kept the Magus.

Thoth Incarnates

By 1660, two generations had passed since Casaubon had disposed hermetic thinking, and during that time those seeking to free themselves from both the magical and the religious prejudices of the Renaissance[148] had found the courage to consolidate their position. These men were the logical next phase in the journey after the pure mathematics of the ancient Greeks, which had been revived in the schools of the Renaissance, and the seismic shift that had placed the sun at the centre of our galaxy and so demoted the religious position of man as the centre of creation. This

---

147 In Australia, we witness the massive ugly coal pits, which make billionaires out of opportunists while adding immeasurably to pollution, global warming, plain ugliness and the destruction of communities.

148 The final witch burnings did not happen until 1679, although the public atmosphere that would tolerate these outrages had well and truly thinned by then.

last shift of course opened new possibilities for asking questions about the creation – but as object. As stated, the occultists had created the ferment and so the will to do so and thus for the first time a barrier was crossed: the sciences were *applied*.

In 1663, a network of new scientists set in place the first charter of 'The Royal Society for the Improvement of Natural Knowledge',[149] a pioneering institution in England dedicated to the exchange of ideas and evidence in a rational environment. Many more such societies followed all over Europe and the growing Western world. It seems that even as the mythical Great Man fell, the magical Thoth was being truly incarnated as Team Great Man but devoid now of any cosmic context.

It was in this manner that astrology passed onto the margins of society. Of course, there is nothing wrong with being on the margins of society – in fact, there are some advantages. The danger is that the astrologer (or any other practitioner in a demoted category) will internalise the collective voice of the new establishment and become trivial and apologetic in pursuit of the art and, in this manner, dis-empower a vital function of mind, the opening up of which is *the* gift of astrological practice. Such an astrologer will not come to know the initiation of mind that sets the program of the Angel Body before him or her and hence before the client/student who may be ready to strive.

In the following chapters, there will be a brief outline, with examples, of the abandoned mythopoetic approach to scientific fact. Then we shall apply this as one of various approaches to reading the astrological map; and we will pass from this onto three chapters on the outer planets taken as major force-energies assisting in the evolution of the soul.

---

149 Interestingly, the original name proposed for the Society was 'The Invisible College for the Promoting of Physico-Mathematical Learning'. Why 'invisible'? Were there echoes of a magical brotherhood, a persistent trace of a longing for an enchantment to accompany their discoveries, now incompatible with the incoming era? Again, perhaps the word found its way in via the two men who founded the Society, both of whom were Freemasons. Perhaps they shuddered before their colleagues as they wrote the words 'Invisible College' into the title, and then deleted them?

THE OTHER SCIENCE
Chapter Eleven

Astrologers call it the belief in catasterism or 'becoming a star'. It was felt by many that the souls of the dead, instead of shuffling and whispering unhappily in the dark underworld (as the early Greeks and other peoples thought), would leave their bodies and float upwards. Where would they go then? They would become immortal and be transformed into stars.

– Jan Kurrels

The Astrologer's Mind

We have now reached the moment when we need to enter the enchanted place, the mystical forest that is the astrologer's mind and to see what we might find there. We will start with the practice itself because what we practise is what we become.

Astrologers converse with other human beings, one-to-one; at the peak of one's practice this may take place almost on a daily basis. The conversations are intense and, as time is limited, the questioning by the astrologer needs to be to the point. The first question is usually 'why have you come?' At this initial moment, the client will tell the astrologer a *story*.

The astrologer will have the astrological map or chart of the client in front of them both and will need to translate the story of the client into its principles and then further into primaries that can be recognised as the action of particular planets. The astrologer will add some transits into the brew, mix in a pinch of sound intuition, the sort that perceives the level of understanding that may already grace the client (for a chart cannot predict the level of wakefulness or self-work that the client may bring into the room), give it all a stir and then reflect back to the client another story, one about the inner nature of her or his dilemma.

Therefore, it could be said that the first gift acquired by the astrologer is one of *translation*, of moving between languages that have entirely different structures. The initial language is the common speech; the other is a relationship between archetype and geometry that is the equivalent of the syntactical component in any language.

The geometry in astrology is chiefly one of the angles between sun, moon and planets and this is initially calculated as the number of degrees separating each pair. However, in dealing with number the astrologer learns to pay attention not only to numerical count but to the *qualities* signified by such numbers, to their ancient meaning. The numbers of harmony, of paradox, of seeking pattern, of mystical fascination enter consideration of the chart. Number quality is essential in considering aspects between planets. It is also essential in the consideration of harmonic charts: these are the charts which result from the multiplication of the base chart by a specific number, giving an entire map of how that numerical quality expresses itself in the life of the person who possesses the map.

The crisp writing of astrology is made up of a series of glyphs. Each glyph is a fundamental, irreducible quality – as, it is said, are Truth, Beauty and Goodness in human potential – and can manifest as either a type of action or event, or as a type of person or object (natural or man-made). In other words, the glyph can represent a wave or a particle, the quality of energy or of matter. The astrologer has to learn to recognise the presence of these qualities in life-on-Earth and to extract their essences from the bedrock of human stories.

Therefore, the astrologer may have under consideration several layers of possibility emanating from the glyphs and chart geometry plus additional layers imposed by harmonic frequencies, and all translated from the concrete problems of the clients. The constant working in this manner results in an unusual sensitivity to *abstracts-in-combination*; however unlike a mathematician who is working with a vast complex of number within rules such as square roots and algebra, what separates the astrologer is that he or she is also working with quality and story while maintaining proficiency in number, geometry and astronomy. It is this recombining of what the modern mind has rendered as 'oranges and apples', as different kinds and categories, that gives the astrologer's mind its particular – some would say its peculiar – perceptions.

The layering on of qualities and abstracts can result also in the opening up of a temporary gap in the process of reasoning. It is a *hesitation outside time*[150], rather like the pause in the trapeze artist's flight when he or she performs tricks. Into that hesitation gap an objective symbol may appear – not an allegorical symbol which can only express one thing as being like another – but a symbol which directly communicates the reality of the mystery it veils. This mental pathway to objective symbols is opened up by an ongoing astrological practice. This process or inner work toward meaning can be applied to events in the history of science itself and without altering the facts.

At this point, we will re-tell two stories of science.

Kepler's Eye

Every author writing on the birth of modern Western science and particularly about the corrections that were needed to turn astronomy into an accurate branch of science will eventually have to write about Johannes Kepler (1571-1630), who was the founder of modern astronomy in the Western world.

There have been a huge number of chapters written on Kepler: in

---

150 There is a website that gives the astrology charts of many of the major astrologers publishing in English from the 20th century. In every case bar one (Charles Carter), there is a tight major aspect between the Moon and Uranus – squares, trines and oppositions. It seems that an astrologer needs the presence of Uranus to cut into lunar repetition, halt its potential for sleep and enable Revelation.

Johannes Kepler: Born 27ᵗʰ December 1571 at 2.37 p.m.
at Weil de Stadt, Germany.

books on music, on optics, on metaphysics, on Plato and the Greek legacy, even a novel by John Banville celebrates the achievement of this
humble person, who struggled all his life and died en route to Regens-
burg where he planned to beg the Emperor and nobles for part of the
salary that had been promised to him and which he, poor and ill, desperately needed. Poor Kepler.

Kepler was an astrologer and a good one; he ran a popular astrology
business in his limited spare time. Kepler's records show that he had
drawn up a chart for his time of conception as well as his natal chart,
calculating his conception as having occurred on '16 May, A.D. 1571

at 4.37a.m.', his mother's pregnancy lasting '224 days, 9 hours and 53 minutes'!

Kepler's birth chart reveals him as a bringer of Uranus in its applied aspect. Kepler has Uranus conjoined to his Sun and Venus in Capricorn and both are impacted by Mars in the sign of scientific harmony, Libra; the Moon in Capricorn is widely conjunct Uranus as well. Kepler's Mercury (data and pattern-seeking) sits opposite the infinite, archetypal Neptune and both are stimulated by square to the obsessive planet, Pluto. Kepler's mind must have been a minefield of transmissions from the two mass destiny planets, Neptune and Pluto; no wonder he found his way into a mentally exhausting career and became obsessed by the aura and mystery of it!

Kepler was the last public figure in Western science to possess the astrologer's mind in its medieval form and (from our point of view) it is everywhere apparent in his copious writings. Kepler knew the earth as a single organism with a soul and he reasoned that, as God had founded our beloved earth on mathematical principles, so any new soul entering the world must carry also its own mathematical form and retain that form throughout his or her life.

Kepler worked for the great astronomer Tycho Brahe under the patronage of the Emperor Rudolph II, pouring over Brahe's meticulously recorded observations of the positions of the planets and trying to discover a pattern in their movements, particularly of Mars whose gyrations were an ongoing puzzle to astronomers. Kepler would spend exhausting night hours at the telescope with Brahe's lists of positions; then he would go off and read horoscopes and listen to the stories of individuals, following which, and having spent hours translating from story to symbol, he would return to the dogged work of the statistician.

As a scientist, Kepler aimed to locate the way in which geometry fused with physics and with mathematics. As a Pythagorean, he recognised that geometry underpinned all of nature. Yet Kepler imagined the surface of Jupiter to be 'like ruby'; Mercury was a sapphire while Saturn – Kepler's own ruling planet – was 'white and dry'.

If one had – as he did – a mind that looked at a fruit or a flower and saw a five-sided geometrical form that was a gift from God manifesting as beauty, it follows that such a mind would have *no fear of accurate science* as such would be a measure of divine truth. Kepler loathed the magical diagrams drawn up by the magician attempting to penetrate the structure of the spiritual world and condemned them as having no basis in reality, as misleading and plain wrong. Kepler believed that accurate science would further reveal and celebrate the Glory of God. Truth and Beauty had combined in his mind.

It was late in the Renaissance when Kepler, using the orbit of Mars, finally proved that the planets proceed around the sun on elliptical paths, not on the perfect circular paths which until now had always been assumed to be the case. Kepler's law of 1609 states: 'the orbit of every planet is an ellipse with the sun at one focus'.

Copernicus with his sun-centred universe, which had displaced the central place of man-on-earth in favour of the Sun, had been disturbing enough to the church hierarchy. Now it was the perfection of the circle itself that was being removed as the natural orbit of mankind! Man is *not* central to God's creation and man does *not* move along a (non-existent) path of perfection. Those of a religious mind felt that their story of God and mankind was being corrupted.

However, there may be another story if we but apply pure symbol images.

Consider, if you will, the ellipse; then consider the circle. If the circle is placed over and inside the ellipse, it will produce an abstract of the open eye. *The open eye is a core symbol of wakefulness.* We wake up physically, we open our physical eyes. We wake up psychologically, we become conscious.

Kepler wrote out his law in 1609, only a few months before Galilei Galileo built his telescope which could be said also to be an 'eye' of the world, albeit a technical one. Kepler wrote to Galileo: they were both Copernicans and furthermore Kepler knew that Galileo needed help, that his accurate scientific observations were ruffling official feathers.

Galileo never mentions astrology. His role in the manifestation of the Eye of Wakefulness was as inventor and observer – his was the *physical end of the vertical axis*. Kepler's was the metaphysical end. It was, after all, the year of the archetypal Neptune square Uranus and Saturn in Aquarius and it was these combined positions in the zodiac that created the spiritual conditions for a scientific shift, should any on earth be prepared to receive the call. Both Kepler and Galileo were ready for the energy, as the high Saturnians always are. Kepler was Capricorn and Galileo Aquarius.

Kepler's Eye is an event outside time and Kepler's ellipse came before the world because Kepler's mind was one worthy of his appointed task. The open eye transmitted out from the gap-world of prime symbols and, like the beam of light refracting as it enters the atmosphere of the Earth, the eye became the telescope, which in turn began to brood upward on the breathtaking physical infinity that is our universe. Galileo saw the Milky Way, he saw four moons of Jupiter…

Therefore, the Eye was opened for humankind and its opening was as a Garden of Eden moment in the thought bubble of modern science. There would be no turning back. Of course, the tragedy is that Kepler always affirmed in his heart that accurate science must be a mirror of God's living laws. He loved astrology, believed in the harmonies of earth, man and spirit and never stopped looking for them as laws. He even gave each planet a musical notation[151] and the Pythagorean term Music of the Spheres would be applied to his work. It is a tragedy to read articles that suggest that we airbrush this vital component of Kepler's mind as an embarrassment which he 'might have thought better of later'. The modern imitators of Great Men can never grasp that the visions granted to those who are ready are received from the Glory and from a hesitation in time – that there may be another story.

Oh yes – and after the opening of Kepler's Eye, Kepler himself became the founder of modern optics.

---

151 The movement of the earth is a repeated semi-tone.

Jupiter the Storyteller

Stories. I have a friend who ran a course in the writing school of a major institution, which he called Story Worlds. The Bible is a story world; science is a story world; astrology is a story world. Each story world acts as a lens for meaning, a reference point, a way of constructing life that is internally consistent. They contain philosophical language tools and none alone should be called 'reality'.

In astrology, Jupiter is the storyteller, the planet that rules the great human themes: culture, the anecdotal and meaningful aspect of the law, religious story, the sport of nations. Jupiter affirms that one's life is a blessing and a rare opportunity. He asks us to improvise within an abundance. Jupiter is also recognised as a protective force in any chart. The inference that *protection is somehow linked with the possession of a story world* ought not to be missed by the astrologer.

Therefore, let us now tell a story about Jupiter, as a gift for him and his angel, in which we link science with that which he would tell us.

Jupiter's positive role as protector and vigorous role as sportsman extends well outside astrological tradition. For the planet Jupiter 'catches' asteroids for us and changes the paths of comets through the force of his gravity and without him the Earth would have been bombarded and many places destroyed! Sportsman-like, Jupiter swings comets in an arc around the Sun. For example, this happened when the Brooks comet made the mistake of orbiting close to Jupiter in 1886: it had its orbit reduced from 29 to 7 years!

Jupiter is the only planet with enough energy to toss a planetoid right out of our system, as Jupiter weighs two and a half times as much as all the other planets put together and his girth is a thousand times fatter than that of the Earth. He is called The Regulator in courses in astronomy because his size and gravitational power have been major influences in stabilizing the fields of the other planets. As in traditional astrology, so in science Jupiter proves to be a true protective force.

Jupiter proceeds around the Sun with his own regular entourage, a 'team' of twelve asteroids with seven travelling ahead of him and five behind. These are called the Trojans. The distance of the Trojans from Jupiter is always the same as their distance from the Sun, which creates a stable triangular configuration, a moving pattern of completeness.[152]

Visually and audibly, Jupiter is of course magnificent. Scientists report that radio telescopes react more loudly to the noise from Jupiter than from any other source, as Jupiter boasts magnetic storms that extend hundreds of thousands of kilometres into the outer radiation belts, has thousands of vortices and his dark side is a continuous crackle of lightning bolts. The coloured bands of Jupiter consist of ammonia and methane along with imine and hydrazine, giving magical blues and yellows; a minor chemical compound produced by ultraviolet light striking the methane component gives off streaks of red. Such colours are forever moving and changing form on the face of the planet.

In traditional astrology, both colour and sound come under the rule of Jupiter, Jupiter in Sagittarius (particularly the early degrees) commanding colour and Jupiter in Pisces letting in sound.[153] And not only does Jupiter 'rage' and exaggerate his flamboyance: as long ago as eleven hundred BC, the people of Ancient China declared Jupiter to be the 'Celestial Arbiter of Happiness' due to his gyrations. Yes everyone, Jupiter does laugh!

Dream of the Red Spot

Of all the storms and spots that regularly light up the surface of Jupiter, it is of course the cyclonic red spot that has puzzled the scientists and caught the imagination. The spot is an anti-cyclone (one that spins in the

---

152 See Guy Murchie, 'Music of the Spheres': the material universe from atom to quasar, Volume One 'The Macrocosm'. Dover Publications, 1967. Murchie has a powerful 'feel' for our brother and sister planets and his books are still a gentle introduction to science, even though some of the reported facts have been altered by time.

153 Among my own clients, there are several with a prominent Jupiter in Pisces in aspect to personal planets who possess a grand piano.

opposite direction to the rotation of the planet) only twenty-two degrees south of Jupiter's bulging equatorial waistline. It is ovaline and over twenty thousand kilometres in length, and could contain at least two of planet Earth. What is more, the spot goes for little trips across the surface of Jupiter – but always returns home.

We are told that the spot developed between 183 and 348 years ago but it must be at the distant end of this range because the Royal Society experimenter and pragmatist Robert Hooke sighted it in 1664.

Let us pause for a moment and look more closely at the spot as an event that is also an invitation into our story world.

By 1664, what would become modern science was well underway. Astrology, alchemy and the law of correspondences were falling away and the new land of the purged intellect lay before us, with its massive, physically beneficial practical developments and exciting discoveries. However, what was about to be lost was a story continuity grounded in wisdom, or rather our capacity to use the lens of story in a manner that enchants. We were like shoppers purchasing new clean pots but forgetting to transplant into them healthy and established plants.

Perhaps Jupiter was angry at the death of his living wisdom story and of humanity's ancient love affair with religious meaning. If he was, perhaps Jupiter-like he sought to publicise his rage to us. He may have recruited colour and energy reinforcements from Mars, his neighbour the red planet and the ruler of the dynamic forces needed in order to give anger a drive. In a moment of perfect geometry between Jupiter, the Trojans on their triangle and the planet Mars, perhaps there was an outside-time gap, a shift in the movement of symbols and an exchange; and through the crack Jupiter was presented with his red cyclone, like a daytime shadow of Mars the God of War – and the cyclone began to spin 'against the grain' of convention, as it became visible from Earth.

Yet such cyclones commonly cease over short periods of time. Then why is the Red Spot still with us? Scientifically speaking, it seems that the spot is positioned between two jet streams struggling in opposite directions, and this slows down the spin of the spot; then when it slows and loses some of its energy, it draws in heat from above and cold from below, which vertical, gaseous input renews the vigour of its spin.[154] In traditional thought, the vertical axis corresponds to the impact of the realm of spirit.

In conclusion and according to this fresh meditation upon story, we have on the face of Jupiter a rage that spins against the grain and maintains its rage by drawing on the energies of its vertical axis; and like two Cyclopsian eyes, the spot and Mars stare at each other across a belt of asteroids and transmit a signal for those to read who can: *when will this madness end?*

36 Degrees Plus One

One final word about Jupiter. In his orbit around the Sun, he travels at the rate of 37 degrees every year. Now, the number 36 (not 37) is one of those 'perfect number' in meaningful geometry. In astrology, it describes the 36 decanates[155] of the ancients; it is divisible by 12 (signs), by 4 (the elements), by 3 (the cardinal, fixed and mutable modes)[156] and by 9 (3x3, the number of completion and higher harmony, the fruit and that which we take joy in, expressed through the ninth harmonic chart). Therefore, you could say that the number 36 is an exact number and, as such, is a closed number, needing nothing further.

However, there is an ancient Jewish tradition that allows the addition or subtraction of the number One – the number of Unity – and that teaches that such an addition demonstrates 'special relationships'.[157]

---

154 This is the most recent conclusion about the spin of the Red Spot. See the science article reported in the 'Mail online' of 15/11/13 (accessed 14/1/14).

155 Each zodiac sign has three sub-divisions or sub-harmonics of ten degrees each; these were assigned to the three signs of the dominate sign's elemental trinity.

156 These are also related to Brahma, Vishnu and Shiva in the Indian tradition.

157 See Keith Critchlow, op. cit. (note 16). Critchlow is writing in the context of biblical numbers. Apparently the number 37 appears again and again in the numerical breakdown of ancient Jewish texts.

Now, if 36 is already a mathematically exact number, why would this other number be needed to complete it? What is a 'special relationship' and, most importantly, why would we need to add the number of Unity to obtain it? What may this have to do with the cycle of Jupiter, the bringer of meaning, around the light of the Sun? I will leave these questions to your consideration, with the comment that each fact, treated astrologically, may become also an opening, that the study of qualities may bring forth other kinds of questions, then a pondering – until a Mystery appears.

Orbital Notes

When we consider the structure of galaxies in the light of cosmic story, it must be taken as axiomatic that there are no random features in the universe at present ('present' because of course galaxies change over time). We are able to study any prominent feature of our universe in the light of story.

To enable the story to speak to us, we need to abandon temporarily some assumptions. In astrology, as in many other things, we have been hypnotised by two limiting factors. One is the constant association of astrology with fortune telling; the other is giving every planet the name of a of Greek God and so bringing astrology – as so much else – under the Athenocentrism of the intellectual fabric of the Western world.[158]

Therefore let us return to our sky bodies without Greek mythology for a moment. Let us look, for example, at what the orbits – what orbital stories – may have to tell us.

---

158 So much of astrology has been based upon Greek mythology: the planet Pluto is compared to the God Pluto, we study what Pluto the God was in Greece and assign the planet characteristics accordingly. To some extent this is accurate and helpful, one measure and area of research. However, to explain away by cross comparison is to diminish the power of astrology, which assists in the construction of the higher bodies and frees the human from the shafts of history – to omit these vertical possibilities – misses the endpoint of astrology and makes of it a pastime or a servant of our post-nineteenth century obsession with psychology, which often draws on the same Greek archetypes.

It has been noted frequently that the dwarf planet Pluto has a great tilt to her orbit and that this is much like the significant tilt to the orbit of the prominent asteroid Chiron, although the tilt of Pluto is greater. Such orbital inclines mean that a body with such an orbit will spend a greater amount of time transiting one section of the zodiac belt than it will spend transiting the section of sky at the opposite end of its ellipse.

The incline also means that the bodies may pass inside the orbits of other bodies – and out again – *as if they have a task to perform within the orbit of the other planet.* With this in mind, let us look briefly at Pluto and at then at Chiron.

In astrological lore, Pluto is known as the planet of purification and transformation. Her journey around the Sun takes 248 years and during that time she spends about 20 years *inside the path of Neptune.* This means that for 20 years out of every 248, Pluto is travelling between Uranus and Neptune, which are with Pluto the 'mass destiny' planets in our solar system. The last incursion by Pluto into the Uranus-Neptune belt happened across the years spanning January 23rd, 1979 to February 11th, 1999.

Let us consider Uranus and Neptune in combination. Why might the vast plane of their combined energies need the periodic input of Pluto? Now, an unchecked Uranus-Neptune – unchecked in that it never receives any purification and as such is able to collapse into its degenerate form – in all probability would result in a kind of mass human madness. History might sum up a Uranus-Neptune mass madness as 'a period of humanity's delusions of intellectual superiority running amok without boundaries or discernment, as it pogo jumps between ideologies that are in equal part inflexible and plain wrong.'

Therefore, for 20 years Pluto enters the Uranus--Neptune zone and purifies the extremities. Basically, she defrags the cosmos! Why only 20 years? As the perfection of all things has it, this period brackets the twelve years in which she passes through the strong-willed and concentrated sign Scorpio and she is empowered by this transit. One could say that she functions from her essence at this time.

Pluto enters her 20 year period of purification while still in Libra and she was there in 1979 as she made her transition; she entered Scorpio in 1983. In perusing world events across 1979 until 1999, certain general comments might be made. Remember that events such as wars and political upheaval are always happening in some part of our world, unfortunately, and so it is relatively pointless to go nit-picking into the wars of the times. Rather, Pluto is about the mysteries of sex and death, of generation and regeneration and our relationship to these great human themes, and she also marks out territories in the collective psyche for good and for ill.

Libra is the sign of partnerships, peace and dialogue as well as a sign of open enemies. As Pluto in Libra passed across the orbit of Neptune in 1979, certain international efforts toward dialogue began. The keynote of the transit was given by then President Jimmy Carter of the United States (himself a Libran), who stood side by side with the Pope, John Paul II, and called for 'world peace'. The same year, Pope John Paul became the first Pope to visit a communist country and he ended the automatic ex-communication of freemasons. Dialogue and the empowering of justice were in the air. Mother Theresa won the Nobel Prize for Peace. The Voyager discovered that Jupiter too has a ring.

As Pluto continued her journey, the New Age movement entered the Western world. This may seem trivial but it was not so: the central tenet of the New Age was and remains the operation of tools for *self-transformation*, the cleansing of the psyche away from the framework of organised religion and its requirement that one 'believes' in an outside authority. This renewal of individual purification and self-empowerment was a truly mass phenomenon in the West and very much representative of an intense upgrading of Uranus-Neptune concerns.

Then, as Pluto entered Scorpio in 1983, the entire international community became aware of the dangers of the sexually transmitted disease known as AIDS. (AIDS was already present but without international acknowledgement; particularly the Africans were aware of it.) Huge questions were raised regarding the consequences of the way we conduct our sexual relationships, about the difference between free love and choice versus promiscuity, about caring for each other without prejudice, above all about death itself, as the disease cut into young lives and brought death into the present.

Pluto entered Sagittarius at the end of 1995, beginning a period of the rising of shadows associated with our poor understanding of religion and international cultural diversity, before she once more stepped back into her own zone at the end of the solar system in 1999. Difficult and dangerous as these cycles often prove, without them there can be no questioning of ourselves and no health in our spiritual system: we will end with madness, bigotry and disease without purification.

Chiron and the Contract of Humanity

The journey of the Angel of Chiron sees Chiron transiting first across the zone of Saturn-Uranus and then into the truly human arena between Jupiter and Saturn. Chiron passes around the Sun and back to his original place once in 49 to 51 years, of which he spends between five and seven years inside Saturn. Chiron does not have a stable orbit and is affected by the gravity fields of both Uranus and Saturn.[159]

Chiron's aphelion (his furthest point from the Sun) falls just outside the perihelion (closest proximity to the Sun) of Uranus; and from the perspective of the spiritual tales embedded in astrology, it seems that Chiron's passing just outside the inner orbit of Uranus provides the opportunity for him to collect a 'parcel' of purified Uranus-Neptune mass destiny. He then travels the turbulence of the Uranus-Saturn zone, coming closer and closer to Saturn. Chiron's own perihelion lies just outside Saturn's orbit.

Chiron transits with a purified spiritual energy – but he has to get around Saturn!

---

159 All of the 'centaurs' are unstable and will exit our solar system millions of years from now, courtesy of the perturbations of the giant planets.

144 ~ *Astrology: history and purpose*

At the moment of Chiron's approach to Saturn there is an almighty cosmic struggle. Saturn is the ring-pass-not of human life and he insists that anyone or -thing wishing to conjoin him be converted into something of 'substance': otherwise he will not let it through. Saturn insists that the Uranus-Neptune elixir carried by Chiron convert to a rainbow of concrete ideas – ideas that can be *applied*. The *As Above* must pass through the mirror and become the *So Below*: else Saturn will not let us through. Therefore, Chiron is forced into the field of the perturbations of Saturn. In the process of the struggle, the field of Chiron turns sideways to get through the denser space of Saturnine contraction: this is the process that can produce the Weeping Wound with which Chiron is so often associated in astrology. Chiron is carrying something that is 'too much' for ordinary humanity[160] and under the impact of Saturn it manifests as injury.

Thus Chiron 'turns', passes through the narrow crystal doorway that is Saturn and enters into the realm of Saturn-Jupiter, where he travels with his assignation.

Now, Jupiter and Saturn have certain things in common when applied to human culture and growth; and perhaps the most obvious is their mutual association with the law and contracts. Saturn is the letter, Jupiter the moral spirit of the law. Saturn follows precedent, Jupiter allows human variability and story. Together, they define that which we agree to contract – what we agree to as meaningful and the manner in which it will be framed.

Therefore, every fifty years or so, our contracts receive a fresh baptism in the waters of pure collective human potential, imagination and invention. Much of science has been disclosed through the *waking dreams* of collective humanity, which is in part the plane of Uranus-Neptune that grants invention and imagination both. However, the tuning

---

160 In the delightful and well-known little book on masculine psychology 'He', Robert A. Johnson pointed to the wound of the Fisher King in the story of Parsifal, the wound that would not heal, as a mark of the young receiving something that is 'too much, too soon'. See Robert A. Johnson 'HE. Understanding Masculine Psychology', Mills House, Berkeley California 1989.

down of Saturn will mean inevitably that a great part of the result will be also mechanical or unaware. Returning to the theme of the age of the World Wide Web for a moment, we notice that the 1979-1999 period of Pluto's incursion into Uranus-Neptune marked the start of our 'internet dreaming' and that this was confirmed 'below' when Chiron arrived inside Saturn between 1992 and 1999. These were the years when the major businesses that gave us and profited by the new dreaming were established. As remarked, such can drag a majority into a new form of distractedness (and screen fatigue) – but there are also those who possess discernment and these will utilise always the new gifts of our practical contract in their own manner.

Both Pluto and Chiron finished their mission in 1999 – and at the practical level left the world with a truly Uranus-Neptune paradox – a revolution in the form of the invisible![161] The more refined levels of the renewal of our contract will of course not be obvious. Only as traces may our research reveal some subtle events – perhaps events in the arts that are founded upon traditional story, or a discreet teaching offered to those who feel the call, or a gathering of communities in a manner that is particularly graced may occur. These things will not be found in a roll call of on-line historical data. They are Angel Body matters, part of sacred history and unacknowledged as we poor distracted glue ourselves to bad news broadcasts and fall victim to doubt. Meanwhile much good, unheeded, flows from the circle of conscious humanity – those with the Mission Body – through whom the higher Uranus-Neptune is utilised to heal and to avert.

In the end, the realm of Jupiter-Saturn appears to ask us all to utilise again, with Chiron as enabler, the gift of spiritual force and to create something of substance, at whatever level we are able to contract. These are the cycles of Chiron.

---

161 It would be a compelling piece of research to track back to the other exits of Pluto and to see what was accomplished during the nearest incursion of Chiron, when humanity was invited to renew its contract with its own greater possibilities.

Twinkle Twinkle

Finally, let us take a break from astronomy and briefly apply our symbolic sight to chemistry.

Much like the Eye, the Star is a primary symbol in human consciousness. Stars turn up in every story in which longing or aspiration is likewise present, and the geometry of stars, whether five-, six-, seven- or nine-pointed, lies beneath many beauties of the natural world.

Stars also 'appear' as the shapes underpinning what would otherwise seem dry, abstract scientific discoveries. One unusual example of this 'hidden star' factor is the seven-pointed star that can be constructed by linking the atomic numbers of the elements in the Periodic Table that correspond to the seven ancient planets. Specifically, we draw up lines that travel from atomic number to atomic number in ascending order: this gives a moving or 'twinkling' star form. We walk as pilgrims from star point to star point.

The Periodic Table in chemistry is a very beautiful thing. It can, in its totality, be set out on a page in several ways, and one of these ways is in the shape of a tree which is another irreducible, universal symbol. The Tree appears in many stories about our origins, in myth and in religion around the world.

Therefore, in the Periodic Table of the 180 elements that make up our world we have disclosed to us *a tree full of stars*. This is the image we are invited to contemplate as the true ground upon which we walk.

In writing on these symbolic story adventures, I hope that you have received a sampling of the prophetic creativity that is astrology. It is time at last to turn to the horoscope.

READING THE MAP
Chapter Twelve

What perhaps upsets most critics of astrology – whether scientists or theologians – is its particular slant on time, which refuses to side solidly with either the objective/empirical or the theoretical/sacred school of thought, but proposes instead a kind of intersection or interpenetration of the two – of above and below – in a moment of presence, the birth (or conception) chart.

– Rob Baker, PARABOLA Spring 1990, 59.

After so many years, how can one write about the sheer joy of reading an astrological map? The practice of astrology is like a ritual within a sacred circle whose object is to resuscitate memory of the Angel Body; or like a recurring festival, wherein the stories of clients are elevated to the symbolic and their dilemmas to great initiations. The journey to the festival has many stages: psychological astrology, simple rules astrology, karmic astrology, New Age you-choose-it astrology, thousands of stories, human themes… finally, astrology as a handmaiden of the Angel Body emerges.

This chapter explores the nature of time and some approaches to reading the map and introduces the temple astrology that surrounds the Angel Body.

Drama or Karma?

Astrology arises from a moment in time, from our first breath, and is utterly present. However, what follows is a problem in astrology, which proceeds from the present point. Is the map a presentation of what has *passed*, of what is unresolved and accumulated as one's karmic printout; or is it a map of one's potential, of the quality of one's opportunities in this life to be lived, of one's *future*? Perhaps we are back on the wheel with 'tendencies' and will be back on the wheel again if we do not confront and transform these. Then again (as the quote above indicates), the map itself falls both within and outside of time.

Whatever one's point of view, sooner or later the astrologer will have to make a decision about time.

An indicator of the power of time is the way in which the astrologer 'dreams' when preparing for a client; this is when they locate the dramatic stories which are set out by the interplay of the 'characters' in the map. Now, in the selection of stories for our clients, we are working with symbols and as such with many possibilities; yet, often the astrologer is helped by the mystery of time itself, by the strange way that we exist both within structural time (beginnings and endings) and in eternity/origin simultaneously. Astrologers are particularly equipped to inhabit this liminal space. Therefore, in the practical preparation for the client it is very often discovered that an appropriate story or objective symbol has already appeared in one's life that very week, something read or heard or thought about that has made an impact of peculiar resonance, and just *before* the client map is drawn and their journey considered. This is the odd thing about time.

Perhaps an example will clarify. I had a client 'Gretyl' who has a Sun square to her Moon and both in a t-square formation with a fourth house Pluto. True to form, I had that week been thinking on – and for no particular reason – Tsar Nicholas and Tsarina Alexandra of early twentieth century Russia, how they had been murdered along with their family and *buried in a cellar* and that this event had overturned the identity of the country, which entered years of turbulence and territorial conflicts.

When I started to speak to Gretyl about tsarist Russia – of all things – and about burials, my client (who had only one fire planet in her chart) had to stretch to connect her personal world to such a story. It is most important to stretch a client's imagination into the world of story-myth,

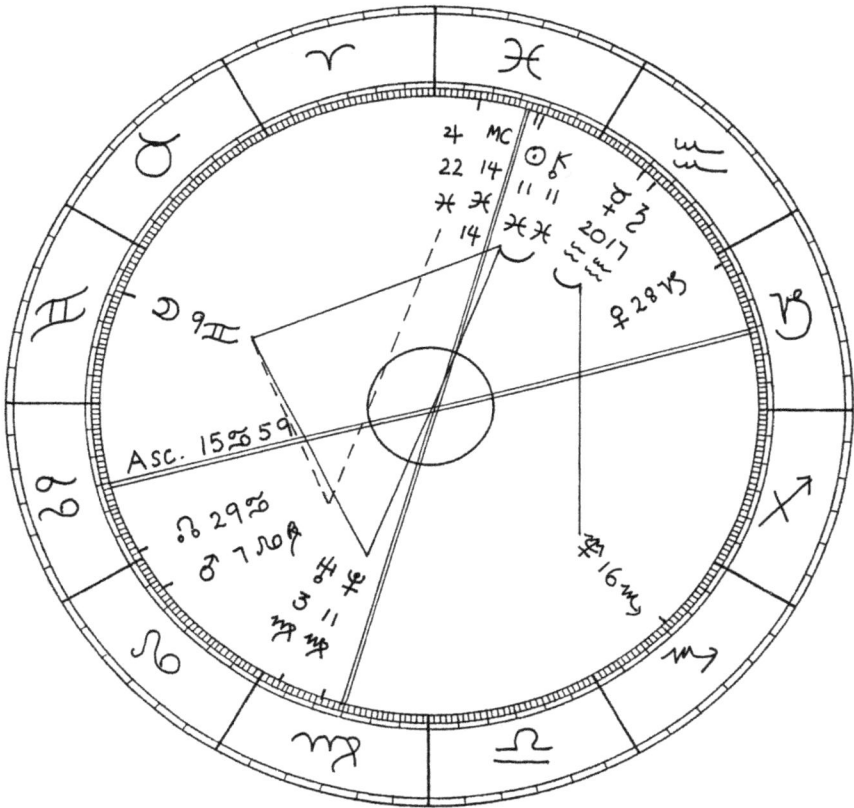

***'Gretyl': Born 2ⁿᵈ March 1963 at 12.20 p.m.
in Colchester, England.***

whatever the point of entry. But then suddenly she said: 'my husband is always saying to me 'oh, you're not *bringing that up again*' and he tells me that I have this compulsion to exhume and overwork subjects that he thought were dead and buried.' It also emerged that Gretyl's previous partner had died slowly and that his dying involved a war of possession between Gretyl and the former wife, which Gretyl neatly described as 'a pissing contest' (!). At this point, our archetypal buried family in Russia was no longer needed, and we worked together to locate and hose down some other 'hidden corpses'.

Of course, one could say that there is karmic possibility here, and perhaps Gretyl had had an association with wartime Russia (or elsewhere) and I myself went through this phase of interpretation. However, these days I have much greater faith in the mystery of time and in the locus of the story 'given'. It is a question of learning to embrace the translogic of the timelines of anticipation.

Staying With the Client

The vast majority of those who visit an astrologer are searching for an extension of the self that does not involve a material view of the world. Many are highly qualified and all intelligent. The astrologer must meet these people on their own ground, and to do this we must have more than astrology to draw upon. We must be concerned with science, medicine, religion and myth and we must trust our instincts in the direction of our studies: we must know when a body of knowledge beckons and what is *within* it that we seek. In other words, we are not looking to study the form of the thing but we look to become receptive to what informs it and gives it life.

It would have been in the late 1980s when a small footnote in a chapter of a barely understood mystical book caught the attention of this astrologer. It said: '... all traditions affirm unanimously that the vital principle is intimately connected with the blood'. I wanted to know in what way there might be a relationship between the physical blood and the principle that is key to keeping oneself spiritually vital. Was the construction of the blood connected to the construction of the soul itself, was there a correspondence? Clearly I needed to make a study of the blood – and so I did, taking copious notes in an old green journal, which was filed away at the end of summer.

Fast-forward five years.

In the early 1990s, a woman – let's call her Peta – began coming for readings. Peta was a senior nursing sister with a Masters in Nursing, and at the time of our first session she was preparing her doctoral thesis.

Peta was the true Aquarius Moon reformer and feminist theoretician. She also had a tight conjunction of Venus and Saturn in Scorpio in her nativity and this was working out in her marriage and in her poor levels of self-esteem across many relationships.

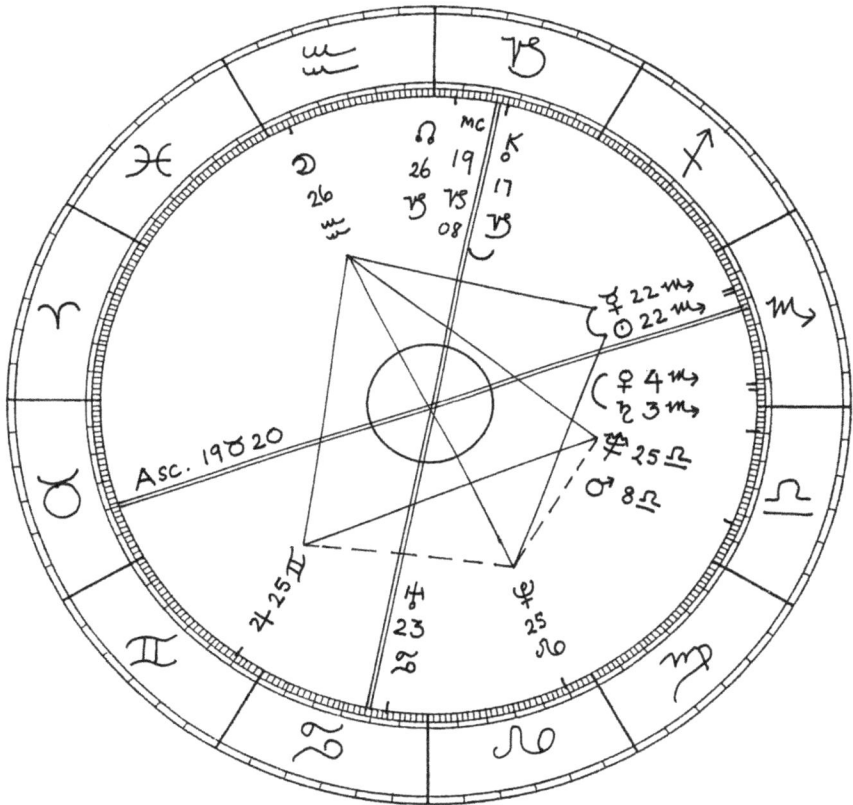

'Peta': Born 14th November 1953 at 4 p.m. in Kendal, England.

In 1998 transiting Neptune entered Aquarius and we saw that it would soon be making a series of squares to her Venus-Saturn. We discussed the possibilities, how to read the meaning of events, set some precautions in place and I gave her the exit date for the transit.

Time passed and Neptune applied and retreated. Then, the day before the final date of the transit – and just (she later told me) as she was musing to herself that 'this Neptune transit has not been so bad and I have enjoyed it' – Peta received a call from her doctor, telling her to get into the hospital immediately: Peta had no white blood cells and might die!

Having had a transfusion and been ordered to quit her demanding job and begin treatment, Peta arrived for an astrology session. You can imagine the look on Peta's face when I produced my old green volume, dusted off and its contents studied in preparation for the reading. This book contained her language but all translated into symbols, which were now applied to illuminate the medical fundamental of her Venus-Saturn, its aspects and midpoints.

Once again, this is a story about enchantment, for when the comment on the blood called out to me, it was certain that it was the Glory who spoke. The comment had an aura. I wanted to know about the blood *in order to* know more about the Glory whose vehicle in us is our soul and its vital principle. The mystery of time is here: for it is always when we approach a subject in its transparency, seeking the light behind it, that we are 'taken up' – like Enoch – and enter into the movement of light that is outside time. In this world, before or after does not count. Perhaps the study discloses a layer that makes final sense of some event that happened *before*, enabling forgiveness or compassion. Oft-times it makes its presence known *after*, as it did in the case of Peta. There is a relationship absolute between the beckoning glow of knowledge sought by the heart, the circularity of time and the small living coincidences that make our lives so beautiful.

As astrologer Angela Voss has written: we seek not the forms of knowledge but what informs them.

The Map as Atmosphere

Let's focus further for a moment on the way in which planetary symbols are read 'downwards' and compressed as concrete characteristics or tendencies of the client. We will take as an example a woman named 'Natalie' and will, for the sake of simplicity and clarity, work with just one aspect. This is the exact square of the Sun conjunct Venus in Virgo with Pluto in the twelfth house and it is found on her Seventh Harmonic chart, which is the chart created by multiplying the original chart by seven.

Let us isolate the Venus square Pluto that is driving her Sun.

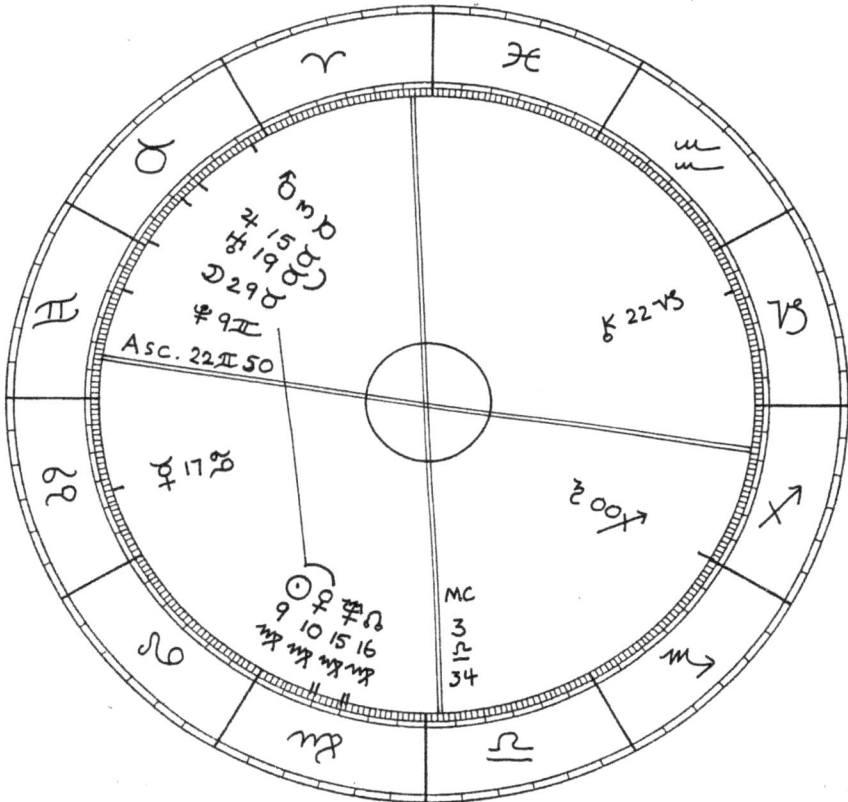

'Natalie': Born 13th April 1965 at 5.55 a.m. in Melbourne. Seventh Harmonic Chart shown.

It has been mentioned that the harmonic charts are maps of the way in which the quality associated with a particular number resonates in the life of an individual. In the case of the seventh harmonic, we have a somewhat mystical map, one that may indicate areas of intrigue or fascination, where a subject contains an aura for the client or compels them in some way; hence, the seventh is strong in individuals who deal in all kinds of *imaginings* – such as the artist, the spiritual philosopher or those with sexual obsessions and disorders.

Therefore, in the case of Natalie, we are looking for *an area of fascination associated with Sun-Venus in Virgo impacted by Pluto* and we need to translate our symbols accordingly.

The first and most obvious conclusion is that here we have someone who is mystically drawn to the erotic: Venus/Goddess Aphrodite meets Pluto, the bringer of fertility and ruler of the organs of regeneration; and yes, it would be (all things considered in the rest of the map and the natal chart) reasonable to assume that for Natalie sex is a mystical experience and/or a compulsion. However, it does nothing for the client to reduce a symbol to one and very obvious manifestation and every astrologer knows that often a client presents with a discreet dilemma and is, without knowing it, searching for another avenue down which to encourage the aspect that is giving her or him trouble. The astrologer must get to the essence of the aspect and then locate other possible examples that may reflect the inner life of the client.

Therefore, Venus-Pluto in Virgo. Venus in her beauty also rules flowers, and in Virgo this might mean tiny flowers on a vine pattern, in which case our underworld twelfth house Pluto would embrace Venus as hidden shadows and as moisture, a heady atmosphere, and Pluto would compress the growth. Putting this together, I asked Natalie whether she liked *moss* and walking on moss in private, shady places, possibly near a stream on a warm day. Her reply was instructive.

Natalie said that only that morning she had been looking at a catalogue of bathrooms. One on offer was a bathroom 'made of moss', where the walls and floor grew a mossy cover and one could step out of the bath onto a mossy floor, which would be maintained by the humidity of the bathroom. Natalie had been fascinated by this and tempted to have such a magical place installed...

One might say that a moist bathroom with walls of moss and steam has a rather similar feel to that of a deeply erotic experience. It has the same *atmosphere*, if you like, the same charge about it. One of the seldom acknowledged abilities of the astrologer is learning to read these atmospheres that always surround planetary combinations, even the most complex of set-ups. The astrologer learns to carry the charge of the combined aspects (or syndromes) into the recesses of Mind – and without prejudice – until the symbols are reconfigured and fused and into the mind a clear example of the combination in action appears. The image-thought is then verified by looking again at the chart. There is always a shift in the charge of one's own thought at this moment, a knowing – 'this is right'.

An astrologer also learns to be sensitive to the moment when the atmosphere in the room 'thickens' because the client him- or herself has said something from the level of soul itself, from essence. This will set the tone of the reading from then on, as the astrologer encourages the spiritual level of the client to speak *its* aims (rather than the voices of fears or idle curiosities having the floor). The reading then becomes an exchange, wherein the astrologer folds the spiritual longing of the client back into the astrological aspects under consideration so that each aspect is raised to its divine potential and pathways are sought in the life whereby that potential may be eventually gained.

A reading requires that we are always awake to atmosphere: we must be able to sniff out the various charges which accompany, for example, half-truths or statements of soul; and nowhere is this more apparent than in the confrontation with our map and the stories we tell ourselves about ourselves.

This brings us to the facet of astrology that relates directly to the awakening of the Angel Body.

The Map as Sleep

All major religions agree that the human race is asleep or unconscious. Awake! Awake! says the Bible; the Buddhists seek a mind cleansed of sleep to enable light. In the early 20th century, the work teacher G. I. Gurdjieff urged his students to trial their state of sleep. For instance, become present for a moment and then see how long it takes before the mind begins playing its old tapes again; or set a task – such as that diet! – and see how long you last before 'forgetting' and before old justifications set in. Of course we are asleep.

In his 2014 address to the Astrological Association of Great Britain, Professor Richard Tarnas remarked upon the shadow side of astrological mapping. He said:

We might... recognise an aspect of astrology's own shadow, that needs to explain and explain away everything within its own particular conceptual framework ('Oh, that of course is explained by your Chiron-South Node midpoint square your Scorpio Vesta intercepted in the twelfth house').

It is this kind of reductionism that seals off the glorious possibilities inherent in astrology, much like a Saturnine gaoler slamming shut the door of our cage and bolting it tight. The map becomes our sleep and our excuse. We lose our vertical possibilities, insight and instincts and the Angel Body recedes.

Yet surely we have map characteristics, don't we? Do we have these characteristics or do we not have them?

The Fourth Way teacher Dr Philip Groves had a Sun in Libra and he once remarked to his class that it was the Libran characteristics against which he had to struggle in order to free himself. He had to work against the tendency to compromise, to have peace at any price, to perennial negotiations...

The planets and the zodiac are signifiers of sleep – and this unconsciousness is our *cage*. In one sense, our astrology chart gives us knowledge of the *personal form of our sleep*, which means – of course and paradoxically – it can also provide critical clues to the process of waking up. Becoming aware of that which we tend to do and be, on 'automatic pilot', is invaluable to anyone who is serious about spiritual development.

Let us take, for example, a map with the planet Jupiter in a tight square to Uranus. Now, Jupiter is a planet that affirms for us that life is our bounty, and when we translate this into common activity, it is Jupiter that indicates where may lie successes in life, where our story might interface with its fulfilment. On the other hand, Uranus will always (for example) be intellectually suspicious of the values that lurk behind the word 'success' and may, when in conflict with Jupiter, find oh so principled reasons to abandon Jupiter. This astrologer has found repeatedly people with this conflict inclined to remove themselves from the arena of their own fulfilled opportunities – citing principled reasons – often at the very time when success appears on the horizon. This is their SLEEP and principles have nothing to do with it!

To awaken, such a person needs to learn to recognise the signs as they appear in the mind: threats to quit mid-opportunity, suddenly getting interested in an exciting and unrelated new direction; becoming 'difficult' and judgmental of others who inhabit a charmed circle, and accept the bonuses that society offers... there needs to be a voice that shouts 'Stop! You know where this will lead – to a rush of self-righteousness followed by repeating the pattern when the next exciting and unexpected opportunity appears' (as it always does for this tribe).

The recognition of and the struggle against the most unhelpful aspects of character is, of course, the initial step toward being helped to construct the Angel Body. 'Helped' because there comes a time when, instead of being agents of sleep, the planetary angels become friends and colleagues in the Great Battle to reverse the downwards spin of the spheres and the resulting unconsciousness of the human being; they assists one's upwards trajectory.

It is time to speak about Temple Astrology.

The Map as Temple

In the ancient world, the selection of the site for a temple structure involved separating off an area from any other use and setting it aside as sacred ground alone. The objective of such a circle of sacred ground was to maintain within the community an ongoing and pure contact with Spirit and its archetypal emissaries. The archetypal emissaries flowed through the planets, with Spirit itself translated below as the Astronomer Priest who was shaper of and witness to their mysteries in the temple precinct.

The floor of the temple ground was designed around whole and spiritually resonant numbers: predominantly 5, 6, 7 and 9. These created a stable, harmonic attractor for the qualities and, as architect Sir Keith Critchlow has indicated, such stability in turn helps 'stabilise the world'. The astrological map, with its identical foundation in sacred circularity, whole number harmonics and archetypal presences, can be said to be – *is* – our temple.

In the time of Jesus and before, one of the seven sacraments was called unction or anointing, when the entire body was covered with olive oil. The oil was said to be the medium through which the anointed individual may join with his or her heavenly image which was said to await him or her, a holy double already set aside at birth. Unction gave an experience of one's angel-in-waiting. It was a temple ritual on sacred ground.

As the rot of second century Christianity set in, unction became the object of (probably sexual) nervousness on the part of the patriarchy: that is why we don't hear about it or its importance anymore.[162] What is important here is the acknowledgement of a possible and subtle interface between the prepared person and his or her Angel. We have already

---

162   In the time of Jesus, it was Mary Magdalene who was responsible for anointing and, as night follows day, it was Mary Magdalene who was recast as a reformed 'prostitute' during the first millennium and excised from her true apostolic position. See chapter on Neptune.

spoken of our side of the contract, the effort or spiritual 'perspiration' that is needed to negotiate the spiritual world. In a way, the perspiration of effort becomes our oil.

If the map is our temple, then it is also an invitation to us to become its priest, for nothing can exist without Life within it, including our map; and so we are invited to perform our personal temple duties, to refine the chart and polish it well. Above all, this is the means through which our Angel Body may be both developed inside time and, by resonance, draw in our angel-in-waiting to grace us outside time. We are invited to become the Temple Priest of our own potencies, without *becoming* those potencies[163] for we can as well lose our priestly possibilities inside the chart. This is the leading paradox of astrology.

It is said that there is a language through which knowledge is sewn into the Angel Body and that this language is archetypes. Therefore, one can readily see how the practice of interpretation through the limits of pure astrological archetype prepares the ground of the mind for an additional light, for a seeing through that discloses the Eye, the Tree, the Snake, the Star, quite outside the realm of phenomena and the complications associated with the analysis of phenomena. These are not psychological archetypes (although they can be applied this way, particularly in dream analysis); these are objective symbols and irreducible – they exist.

We turn now to the old astrology and to the planets discovered after the telescope and the possible meaning of their discovery.

Ancient Planets, Modern Trinity

As every astrologer knows, in the ancient astrological system each zodiac sign was assigned a planetary ruler in a symmetrical design as follows:

---

163   This would be called 'identifying' with them in the Fourth Way system.

AQUARIUS	Saturn
PISCES	Jupiter
ARIES	Mars
TAURUS	Venus
GEMINI	Mercury
CANCER	**Moon**
LEO	**Sun**
VIRGO	Mercury
LIBRA	Venus
SCORPIO	Mars
SAGITTARIUS	Jupiter
CAPRICORN	Saturn

Notice that the Sun and Moon are placed at the centre of this system and the planets proceed out from it in orderly pairs. We can see the symmetry clearly and the complimentary meanings of the joint rulerships become apparent.

NIGHT	Moon (Cancer)	Sun (Leo)	DAY
RESONANCE			VITAL PRINCIPLE
EMPTINESS			LUMINOSITY
LANGUAGE	Mercury (Gemini)	Mercury (Virgo)	REASON
NATURE	Venus (Taurus)	Venus (Libra)	HARMONY
CONSCIOUS FORCES	Mars (Aries)	Mars (Scorpio)	UNCONSCIOUS FORCES
SOUND	Jupiter (Pisces)	Jupiter (Sagittarius)	COLOUR
MIND (structure of)	Saturn (Aquarius)	Saturn (Capricorn)	TIME-SPACE (structure of)

Uranus was discovered in 1781. At that time of course astrologers could not have known that there were two further planets awaiting discovery. This was also the period in which science as we know it was being born and many traditions were under siege; and so astrologers – no doubt maintaining their own traditional system – assigned Uranus to the sign Aquarius, thus making Saturn secondary.

However, now we know that there are three outer bodies orbiting the Sun and only three. So we need to look for a further symmetry in the plan of the skies.

The outer three have been explained over time as being the 'higher octaves' of the personal planets Mercury, Venus and Mars. They represent the qualities of those planets raised to a greater frequency and thus to an evolved and refined manifestation of those qualities. Such an assignation of Uranus, Neptune and Pluto as carriers of a higher vibration of the personal planets seems to be a true one but incomplete. If we look again at our planetary list, we will see another symmetry, this time beginning with Mars at the centre and proceeding out in both directions, in the reverse order of the outer planets.

	Higher Octave
SUN- MOON	(None)
MERCURY	URANUS
VENUS	NEPTUNE
MARS	PLUTO
JUPITER	NEPTUNE
SATURN	URANUS

Thus Mercury and Saturn, the two planets associated with mind and reason, find their blended higher vibration – their counterpoint, if you like – in Uranus; Venus and Jupiter, associated with worldly blessings, find their higher vibration in the mystical heart of Neptune; and the Martian forces rise to the power of Pluto.

The Sun and Moon remain forever unique, as they strive to blend in a marriage of luminosity and emptiness, which constitutes enlightened being.

The higher octave planets are also known as mass destiny planets – and here lies the clue, for their discovery may have marked a call to *humanity as a whole* to step up its collective vibration during the bridging period before the globalization of market-driven materialism backed by an atheist scientific elite would take hold. I am inclined to think that these discoveries *at this time* represented a release of knowledge, that Uranus, Neptune and Pluto are revealed knowledge of the inner plane, just as much as they are discovered facts of physical history. Perhaps they were revealed because those who were maintaining traditional spiritual thought were going to need some help: they were going to need some clues about their own evolution and a muscular means through which to re-engage with what they already know.

Therefore, the mystery schools that are Uranus, Neptune and Pluto were presented for examination; and astrologers everywhere began to work on them, trusting that they would give up their secrets and perfume the human mind with renewed wonder. This is indeed what they have done for astrologers, these three. Uranus, Neptune and Pluto have shown us the higher octave paths of The Stranger, The Glory and The Vehicle, pathways axiomatic to the construction of the Angel Body – and thus to our greater empowerment to become a radical part of the solution at a critical time.

We are ready now for our summary chapters on the outer planets.

URANUS THE STRANGER
Chapter Thirteen

TRUTH
MIND

SHOCK
REVELATION
METANOIA
ASTONISHMENT

God is never mentioned in the Bible.

— Perez of Barcelona, Madua 1558

I don't believe in astrology. I only practice it because it works.

— Astrologer with a prominent Uranus

My moments of despair teach me that my life, as I have conceived it, is absurd. But what is beyond despair's ken is the possibility that it is not life itself but my whole ordinary way of living that is absurd.

— Jerome A. Miller

The Planet

Uranus was discovered by William Herschel in 1781, when he noticed an unusual 'comet' among the stars of Gemini – Gemini, the mercurial lower octave of Uranus, the communicator, the sign that brings news.

There is some dispute about the naming of the planet. Some say it is named for Urania, the muse of astronomy but most accept that it was named for Uranus, the Greek god who was the first ruler of the universe and the personification of the sky.[164]

Uranus rotates on his axis at virtually right angles to all the other planets – a signifier of his cosmic character if ever there was one! This means that the axis lies almost on the plane of the solar system itself. Such an angle gives the planet extreme seasons of light and dark spanning up to 40 years.

Uranus does have his thought-rings but these are faint and were only discovered in 1997, during Pluto's first recorded incursion between Uranus and Neptune, when it was still in the sign Sagittarius, the sign of growth in knowledge. He is the third largest planet, a 'gas giant' of a blue-green hue and is 2,870 million kilometres from the sun. Uranus has seventeen moons.

Uranus and Presence

The position of Uranus in the chart will indicate where we are striving to become conscious and the role of any transit of Uranus is to wake us up – this is the beginning and the end of it. What happens under Uranus will depend upon the extent to which we are already awake as human beings or *willing* to awaken. Uranus will use any skilful means at hand to wake us up, to break our astrological patterns. As such, it represents an opening to the change of Mind necessary to any sincere intent to develop the Angel Body.[165]

---

164 It is interesting to note here also that the Greek word OURANOS can signify 'that which covers', 'that which conceals' and 'that which is hidden'. The latter meaning makes reference to the blue veil of the universal Mother and, if applied to the visible sky, indicates astrology's capacity to hide *or* to reveal the great truths.

165 Of course, the 'opening' may close again. The action of Pluto is necessary if we are to retain the insights of Uranus.

One client of mine, a Capricorn Ascendant with a tight T-square involving the Moon, Mars and Saturn, suddenly won two awards in one week when Uranus passed over her Gemini the Twins Sun sign. This surprise broke her entrenched pattern of assuming that life was 'against' her and that all is struggle and oppression. I myself was repeatedly woken by a fault in my car horn one cold winter, until I really woke up and remembered again my philosophical base and direction and began to rise at 3a.m. and study. This was during a transit of Uranus over Jupiter.

However, Uranus can be savage if one is asleep. One student was struck by a car and injured her right hand during a Uranus transit. As it happened, she had finished her final exams the day before, yet she was bicycling back to school along the same route. Why? Habit – lack of consciousness.

Of the many gifts of Uranus, perhaps it is wakeful Presence that is most treasured by the soul.

Uranus has much to do with the bestowal of a talent that is uniquely ours. In the realization of our individual destiny it is our *talent* that is our own particular ally and which makes all of us 'outsiders'. Moreover, Uranus tells us that, should we pursue the line of our difference, of our own gift, we will position ourselves in a manner that will make progressive revelations possible. One's talent is the first midwife to the newborn Uranian pilgrim.

Often the direction that natal Uranus wishes to take is revealed during the first transit of Uranus square Uranus (sometime between 18 and 22 years of age). This is a period of rebellion and disturbance, sudden events and shaking up of the life; and within the disruption and as long as we are doing something completely fresh or associating with new people or doing the usual in an unusual and new manner – well, in the opening created by these interventions a message from higher Uranus may appear. There is a mental excitement, a new way of thinking, a light in the mind.

You see, at core, the first Uranus to Uranus transit contains the seed of a great possibility, for it is first and foremost *an act of rebellion against oneself.* It is the first major opportunity we have to delete the fate we have been handed and to choose our own course, even if the choices are being made inside a sea of instability and tension, which seems to us to require some grand act of perversity. Uranus brings a bolt of lightning and a charge, an excitement born of the potential of the human mind. This is why we often 'discover' something or suffer a (sometimes unpleasant) surgical correction to a course we were taking that could only have led to mediocrity, even if that mediocrity may have been housed as social success. It is the time when we have our first opportunity to stand our ground and be who we are; and if we catch a glimpse of what it might be like to be ALIVE at this time, Uranus will use that pause, however miniscule, to deliver his gift.

Yet in our society, few recognise that this is a time of initiation and so we also get wrong-headedness, lateral jumps, 'yes but' arguments going nowhere… it is all an energy with a superior potential, hovering unused behind manifestations of the random and perverse.

Celestial Physics

Each planet is a mystery school. It contains an unspoken essence which opens out into primary qualities and further down the scale, into manifestations of those qualities, into the categories over which the planet has rulership. The primary qualities can be expressed in terms of their greater or lesser degrees. For instance, as we have noted in the case of Uranus, revelation is a primary quality of the planet and axiomatic to its essence – but shock and disruption, also associated with Uranus. Shock is the same energy but of lesser degree; and astonishment is of greater degree than revelation.

Now, if we consider the relationship between the themes that describe the essence of a planet, it is vital that we use words of the same degree, words that are in 'community'. For instance, emanations of Neptune include both sacrifice and betrayal: these words indicate *themes that can be explored together.* On the other hand, revelation and shock are not of the same degree: one is a diminution of the other.

Planetary emanations of like degree are in community and this community constitutes its mystery school. All who enter the school will come to know that there is much hidden there, particularly in the relationships between qualities of identical degree. Therefore, what is the relationship between sacrifice and betrayal? Between revelation and presence? This is the study that we call pure celestial physics.[166] It is a mystery of Uranus and food for the intuitive mind. Eventually, every astrologer must enter this realm.

Saturn in its Aquarian aspect blesses the collective mind of humanity with an innate receptivity to the fundamental structures of celestial physics, and it is Saturn's higher vibration, Uranus, that lights up our latent knowledge through moments of intuition and revelation both. Saturn itself, wherever it is placed, has to do with the primary principle of inversion or reversal of the natural order. Saturn materialises the spiritual and thereby both obstructs *and* materially manifests the intentions of spirit. Saturn will convert wisdom into dogma, harmony into bondage...

Core principles need to be contemplated when we are making major decisions in life: they are great facilitators as they connect upwards our mercurial mental processes to an unwavering truth.

There was a man – let's call him Jonas – who was making a decision about expanding the scope of his work, but he was frightened that he would lose control of his next project should he do so, as it would be too big to control. This man asked his spiritual teacher a simple question. He said: 'In order to take a step in personal development, is it

---

166 There is a branch of knowledge pertaining the inner Islamic studies that is called celestial physics. The subject takes in images of great beauty such as the movement of the spheres through seven pleromas, each guided and positioned by its archangel. If we remove the imagery, the clothing, we see that these signify living principles that both secure and dance the continuum that is man and cosmos. The study of these principles and their various combinations is a function of Uranus in his quest for Truth.

necessary to first make a sacrifice?' The teacher told him that the right sacrifice was indeed a function enabling transformation. Therefore, the man surrendered his inner insistence on control; and in doing so, he discovered a world of greater light where other people of quality appeared and assisted the project in the right way…

Without an intuitive grasp of the universals of sacrifice and transformation and the introduction of these considerations into his decision making, Jonas could not have come into the spiritual condition in which he could be found by his embryonic community of light. He would have simply made mental lists of reasons to do or not to do the new project. Uranus invites our little minds to enter the greater spiritual Mind. Here, we learn to combine and recombine the qualities, the living truths of the planetary language, and to discover their laws: what they produce in combination, what might happen (for example) if one were to combine the transformed face of one primary with the inverted manifestation of another? The concentration on such things when looking at a chart paves the way for the Angel Mind that thinks in archetypes.

Shock, Revelation, Metanoia

When we look back on our lives, it is the things that came as heightened experiences which we recall because we were present to them. They have become the things we treasure. The through line of these moments of awareness is the true story of one's life.

It is said in traditional thought that Heaven is ever 'inward', a paradise of extended perception free of attachment to the tokens of outward life. Uranus not only turns us upside down, he turns us inside out.

When we are unconscious, Uranus will always come as pure variation, a sudden change of pattern, such as rule breaks, exposures of that which we have refused to face (family secrets), tensions and explosive releases or all manner of gee-whiz goodies and unexpected escapes. However, for those who glean the sparks of light left in the wake of Uranian passes, the exposures and variations will begin to fold into periods of true revelation. We ponder what happens and gradually something is *given* that is beyond mere working out. We go so far in our own line of thought, then suddenly our range is extended for us by the addition of another truth.

Things do not so much 'make sense' as they make sense *in another way*. We are not so much displaced by Uranus, rather *our existing displacement is corrected by intervention*.

Glean further and we arrive at the door opening onto metanoia.

The word 'metanoia' means 'change of mind'. However, it is not changing one's mind as ordinarily understood nor is it learning from experience and growing wise. Metanoia is the process whereby we achieve the connection between personal thought and Universal Mind, thereby reversing the very manner of our thinking. The 'so below' becomes the 'as above' and there is an achievement of a certain and steady mind full of spiritual wonders.

The road to metanoia is signposted by shocks and revelations. How suitably Uranian, how paradoxical it is that the road to the steady mind is signposted by shock.

It is time to look at a case study.

Case Study: Static to Enlightenment

Benjamin is a client and friend I have known for many years and, being well into his sixties, is old enough for us to see the shape of his life.

Ben's is a difficult chart: he possesses the triple conjunction of Mars with Saturn and Pluto common to many born after the Second World War; these energies were activated by a stressful relationship with his father, who was a bully. Partly in consequence of this, Ben gradually rejected the aggression option as the solution to any problem, choosing instead to fulfil the potential of his Rising Neptune in Libra. Ben embraced Buddhism and made a career in the world of film and stage makeup and mask, working with the aesthetics (Libra) of illusion (Neptune). Such a profession is reinforced by his seventh house Moon-Venus

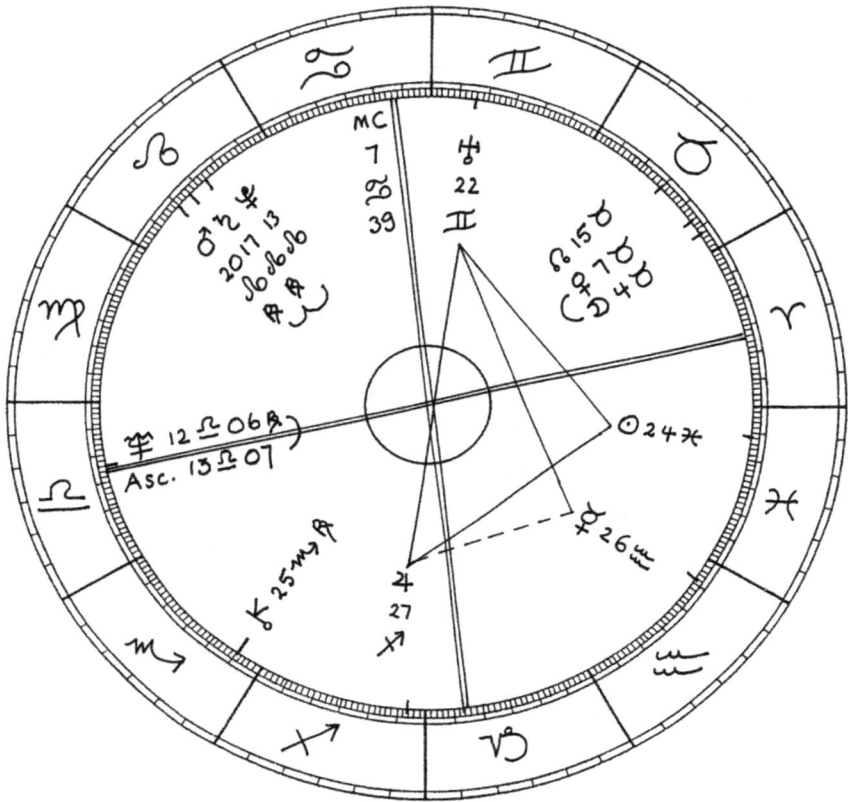

'Benjamin': born 14th March 1948 at 7.30 p.m. at Kerang, Victoria, Australia.

conjunction in Taurus, which has him continually touching faces and drawing out the beauty in the other. Further, Ben has a long-time, successful marriage to a woman with the Sun in Taurus.

However, for the purpose of this chapter, it is the prominent position of Uranus – elevated, square to the Sun, opposite Jupiter and trine to Mercury (conjunct the asteroid Juno) in Aquarius – that we will examine.

The young Benjamin was a combination of rigidity, oversensitivity and intellectual dominance, albeit in a charming and slightly obsequious manner. He carried with him an enormous amount of static, whether as debater or entertainer of guests, and he tended to polarise others: one was a friend or one was an enemy. He had that Mercury trine Uranus canny ability the think on his feet and argue in a devastatingly witty manner (even if he was wrong). He made all the classic Uranus opposite Jupiter attacks: resented the successful, jumped to negative conclusions when generosity was required, would not allow the humanity of others if they were 'wrong', nor offer charity, and he smelt the rats of political incorrectness everywhere. (An artist friend has painted a portrait of Ben. In it, he has a rat at his shoulder.)

Ben conflicted with academic authorities and finally, in true Uranian style, abandoned his degree and went freelance.

By the time he reached his late thirties, Ben had done a massive amount of brilliant work in conceptual design and realization of ideas for performers in the stage, film and music industry. However, he had also made enemies and participated in more than his share of confrontations. (These disasters often occurred when his Mars-Saturn-Pluto broke out in Eleventh House situations. This had an enervating effect. One can't walk through life in a heightened and unresolved state. Something has to give. It did.

Ben was diagnosed with AIDS at the time when it was thought to be a death sentence.

In fact, this was when Ben's life began: the diagnosis proved to be the shock necessary to begin the process of redeeming Uranus.

Having found the ideal enemy at last in his own circumstances, Ben moved to research and understand the nature of the disease, drawing upon his considerable intellectual and spiritual resources. He studied AIDS as a function of Air Prana in Buddhism, put himself into silent retreats where he dealt the death blow to his Mars-Saturn-Pluto anger, then entered the world of medical and mind-body spirit theory and cutting edge neuro-science, all with the grasp and pattern perception of the true Uranus.

Ten years after his diagnosis, Ben had the full AIDS crisis. He lay in a hospital bed in a body weighing a mere 38 kilograms and we were all certain he would die, including the doctors. Then a strange thing happened. Suddenly, Ben sat upright in the bed, opened his eyes and announced: 'You can take me off the morphine now'. He then went back into his coma. What could they do? The doctors took him off the morphine, he slowly recovered, regaining 30 kilograms in the process, and today is thriving and working in a major city.

What had happened? Ben told me that, while he was comatose, he had retreated into a mandala that he had been constructing on the inner plane for years as part of his Buddhist practice. In so doing, he discovered in a waking coma that these practices are 'real', not just something to do mechanically because a teacher told you to or you read it in a book. He was desperate to make me understand this when we next met. He didn't want my 'opinion' or appropriate quotes: he wanted me to WAKE UP AND KNOW IT IS REAL. This was the revelation part of his process.

The process of metanoia is being undertaken slowly, for it is a role of Uranus to know that society itself is 'mad' and full of self-serving blind spots, and to act upon that knowledge: therefore Ben is still a thorn in the side to the establishment and its minions! However, Ben has become steady and discerning in the application of the gift he carries, the gift of exposing lies and challenging corruption. Should one parrot common opinion without due thought in his presence, he will simply say 'that is not true', which invites us all to think further. He still recognises instantly any violation of principles but he is able to let it go[167] in the name of the stillness of mind that comes from self-conquest. He is no longer full of static and has the light of his ready humour turned upon himself as much as the world. He lives with his wife in a Taurus planets' paradise, with garden, soft fabrics, beautiful objects and animals and welcomes 'outsiders' to his table.

---

167 The wonderful astrologer, the late Charles E. O. Carter writes of people with Sun square Uranus: 'they are difficult people to be on the wrong side of, and in such circumstances can be "good haters". Carter: 'Astrological Aspects', L. N. Fowler and Co. Ltd., 1971.

This disciple of Uranus has achieved, helped by shock and revelation, the higher vibration of the planet.

Higher Octave Applications – Using Thought

Uranus is traditionally recognised as the higher octave of Mercury. Now, Mercury in his Gemini aspect has two complementary functions: his descending function makes him the messenger of spirit and the interpreter of the message, which enables the blessings of spirit to enter into language; and his ascending function is as the *guide* who leads the individual through his or her changes from one life cycle to another or *from one state to another.*

Mercury in his Virgo aspect works with systems and patterns: he locates and expands the logical frameworks without which language would be useless. Note that the symbol for Virgo is a mermaid and she is also dual but, in this case, spirit and man combine in synthesis. Mercury as Virgo is cohesive.

Now the function of Mercury as both the guide who may lead us to a change of state *and* as the logical synthesiser is the first riddle associated with the transposition of the way of Mercury into the way of Uranus. For higher states cannot be framed in any logic that we can recognise. In the world of Uranus, everything that we normally see as logical seems now to us quite mad; many comedians have a prominent Uranus!

I had a client who had previously been in the habit of taking off her clothes in the middle of (for example) a department store. Naturally, the society put her on drugs. Another very 'sane' client with a Sun conjunct Uranus and both square to the Moon is a psychiatric nurse. She loves working among people who march up to her and announce 'I am King George today' or some such: she considers this far more honest (and entertaining) than being out in so-called society where people are pretending to be what they are not anyway, and this is considered 'normal' behaviour…

I'm sure you get her drift.

Therefore, Mercury can only deliver us to the door, as messenger, guide and critically as discerner – for it is the Virgoan face of Mercury that casts out anything half-baked along the road – but Mercury cannot encompass the greater frequencies that will be needed. For Uranus is a brave planet and once its function 'opens' after years of Mercurial discernment, exploration and practice, one is never the same. Something *sees* the irrationality of it all. There is a reversal, a metanoia, and what follows is a revolution in one's life. The statement from the Uranian astrologer and the Jewish mystic at the head of this chapter are examples of the assumptions mocking humorous voice of the planet.

As we enter the realm of Uranus, the rhythm and pulse of our thinking changes and Mind begins searching for another paradigm.

It is common knowledge that the quality of our thoughts produce the relative intensity of our brainwaves; for instance, that anger increases the intensity of the pulse implosively and can produce headaches. The study of brainwaves and techniques for intervention belong to Uranus.[168]

The mind produces waves and energy, fuelled by imagery and repetitious self-talk. Such thoughts appear to develop lives of their own, as if they are circling externally, waiting for one's own random musings to create a point of entry, so that the old tapes may be played once more and the mind become stuck in the unresolved. It really is the case that *our thoughts think us.* This is our gaol. The process suggests the rings of Saturn – a circling of carp-carp-carping bands of creatures around a Saturnine centre of gravity that cannot innovate (although Saturn is a wonderful researcher). In contrast, the finer rings of Uranus around a spirit-blue planet may suggest some other course.

Imagine that you are in the midst of an unresolved dispute with another person. You take a walk and he (the disputant) inevitably pops into your mind and immediately the circling begins – carp, carp, carp,

---

168 There is an enormous amount of research on this. For example, at this writing it is thought that applying contradictory audio rhythms to either hemisphere of the brain may cause the brain harmonics to search for a solution at a higher level. It is said that one outcome of this is that one achieves one-pointedness of mind.

criticise, criticise, 'what I should have said'-ing, self-justification. These thought waves are energies that need to be divested of the accompanying images, which both fuel and paralyse them. The waves need to find instead a neutral and superior metaphor. Perhaps the waves feel like the rhythm of sandpapering, like a pushing relentless roughness. Once we divest our brainwaves of their justifying images and enter the realm of metaphor, we are already in a more refined state, free of the rings of addiction. We have taken the first step on the vertical axis, the initial step in our quest to use our thoughts.[169]

The use of thought is an advanced meditative method and contains a mystery of Uranus.

When we are able to recognise and dis-identify with our own Rings by stripping them of the food of repeated storytelling, we also become able to recognise in others that constant circling of their own pack of sirens. We see that all disputes and fake interactions arise when one's present thought identifications enter into bondage with those of another.

Once when travelling in Australia, I happened to see a cheerful sign in a small town offering massage. As my shoulders were sore, I stopped and went inside the shop. The masseur appeared: middle aged, dressed in casual colourful clothes and with a ring in his ear. He introduced himself as 'Crazy'. Of course, I assumed he must be a hippie throwback struggling to live outside the system. I could not have been more wrong! As it turned out, this man had worked successfully in the corporate world most of his life until one day he recognised that all of the

---

169 This is only the initial step. In the book 'Going Home, the journey of a travelling man', author and Fourth Way teacher Reshad Feild describes a man who later became the much-needed chef for his fledgling spiritual group. This man, unknown to Feild at the time, was in meditation halfway across the world in Afghanistan, when he 'received' a thought telling him that there was an Englishman in Los Angeles who needed a cook! The man immediately went to L.A. without knowing who needed him. There, after three weeks of searching and courtesy of a random knock on a random door, he managed to locate the source of the thought waves that had travelled to him outside time and space. See Reshad Feild: 'Going Home, the journey of a travelling man', Element Books Limited, 1996.

relationships he had himself and observed around him were 'fundamentally dishonest'. As a result Crazy left to study samurai techniques in Japan; then he and his wife took the radical step of giving away most of their money (millions of dollars) and seeking a life of honest communication. The little shop with its cheerful massage sign was simply Dizzy's invitation to come in and to tell the truth even if it was for the time of a ten minute neck rub.

The turn in Crazy's life had taken place under Uranus transits. Uranus is a brave planet.

Uranus Asleep – the Ultimate Paradox

Something has to give when we find Uranus unconscious and those in such a state find that their lives are full of dramas, the wrong kinds of sudden reversals and edgy psychological tensions. Here, in brief, are some cases in point.

One of my clients, Daisy, has a tight conjunction of Uranus and the Moon, with both square to Neptune and to Saturn. Daisy has spent so much time in a state of *hypervigilance* (her word) that she has begun to have sudden blackouts, when she just collapses onto the floor. Daisy is a strong feminist, who has been swift to condemn the 'lunar' type of woman – those who see themselves as primarily supportive of a husband, who marry, change their names and make lots of sandwiches for fetes. The consequence of this combination of hypervigilance with the cutting away of what is after all her own potential humility has resulted in breast cancer. She has had, as of this writing, two mastectomies and has had a pacemaker (correcting the Uranian rhythm problems) inserted into her heart...

Uranus asleep: too vigilant, swift to condemn the 'politically incorrect'.

Another client, Patrick, has a heavily aspect Uranus in exact conjunction with his rising degree and no water signs in his map. At the end of our first reading, he pushed his chair back, jumped to his feet and announced: 'This has been a waste of time.' Patrick's trouble had been his inability to comprehend his mother and sisters, all of whom possessed the water sign sensitivities and ability to go into communion with the atmospheric feeling of the other person and shaping the conversation in a way that layers of meaning were present, some spoken, some implied. Patrick just did not know what I was talking about, although he is highly intelligent and open-minded enough to attend an astrologer.

Uranus asleep: rude, blind, psychically 'dead'.

Then there was the actress Katie (Sun opposition Uranus) who began her career by winning a major award in theatre. Yet, 15 years later she was seen walking in traffic carrying her hand-made puppet with which she entered the windows of stationary cars. She had all but become a bag lady and the ultimate, if creative, eccentric. I observed her once on crutches (she had had repeated accidents) insisting that an official give her money. Katie remains one of the fastest, wittiest women I have ever known but this did not help her life, which remained an invitation to reversal, accident, argument and confrontational eccentricities – all sleeping Uranus phenomena.

We have all grown tired of being 'corrected' by the unconscious Uranian: I myself have experienced having my view of what the weather is like today being corrected before I finished the sentence! And we grow tired of the excuses of afflicted Uranus, who cannot be on time because something has always 'come up' or the Mars-Uranus hysterics and car kicking under pressure.

Ego-perversity, dramas, 'I'll do what I like and you can wait' tendencies and 'always right' self-justification. Sigh. Uranus asleep is a trying, trying creature.

Astonishment of Mind

However, if Uranus asleep is the most trying of all the energies, it is in direct proportion to his ultimate possibility, which is the true fulfilment of the human mind.

In chapter three, we spoke about the Chariot 'seen' by one of the prophets of the Exile, Ezekiel; and we spoke about Enoch being raised up and guided by the Archangel Uriel, who gave him a vision of the structure of the heavens. These experiences and the respect accorded them are examples of a visionary awareness all but lost in the West. (These days, Ezekiel and Enoch would join the others of drugs in the psychiatric ward, where my Uranian client would no doubt nurse them...)

The moving chariot with its astronomical construction, wheels within wheels and faces, is considered the master image of a branch of Jewish mysticism called Merkavah Mysticism. Merkavah is the word for chariot. The chariot represents (is) the process of binding the great potential of human awareness to the consciousness of God, during the era when the universe was being founded from the spiritual world. The chariot image is a vortex of divine power, which we regard with awe.

It is said that the fruit of Merkavah Mysticism is the acquiring of a state of astonishment at the Presence of God. It is the capacity to live in permanent astonishment that was sought by the Rabbis, they who participated in a line of succession of Merkavah teachings supported by the heady memorizing of scriptures and meditational practices.

Astonishment is a beautiful word and it is to be found elsewhere in the Western tradition, perhaps most notably in the state that overpowered the disciples of Jesus, when they saw him return after his experience of entering the Glory and speaking to Ezekiel and to Moses (known as The Transfiguration). Those who saw him afterwards were 'astonished'; their mercurial minds were so confused that they became ripe for a revelation of spiritual power and were astonished at its presence.

Presence of God: let us tease out the word 'presence' for a moment. For it is no simple piece of sophistry to point out that one cannot experience God (whatever that may be) without God being present. God cannot *be* without God's Presence. Therefore, it is our own increasing capacity for awareness that may weld our small presence progressively to the great Presence, until the Glory of God opens to our sight.

Furthermore - and not to be too reductionist but this *is* a book on astrology in its mystical context after all! - it is, in the opinion of this astrologer, the continual practice of astrology, with its erection of star-maps in the light of perennial motion and story, that makes of astrology a cousin to the ancient Merkavah Mysticism; and it is the Enochian portal that is Uranus which opens at last into a heaven of Astonish-ment. Then, as we seek to apply our astonished state, we may experi-ence the redeeming paradox of Uranus. We begin to participate in life as lover of *and* as witness to the natural world. We experience union and immanence in the same breath as detachment and transcendence. For both God and Nature wish to be known by us. Being present to the knowledge embedded in Nature – astonished – qualifies us to utilise the angelic mind to add value to her as true scientists, in the state of aston-ishment at the blessing we have received. No longer as fallen angels but with minds full of dynamic chariots and bodies walking through a forest of stars: this is the redemption of Uranus.

Becoming a Stranger

The pre-Socratic philosopher-poet Parmenides drew attention to the state of humankind in the sixth century BC. He wrote:

> ... *for helplessness in their chests is what steers their wandering minds as they're carried along in a daze, deaf and blind at the same time, indistinguishable, undistinguishing crowds.*

Thus does Parmenides describe the path taken by us, as we wander about pursuing phantoms and 'going absolutely nowhere'. It is a tough passage.

The gift of Uranus is the cycle of interventions into our state of 'going nowhere'. We receive a spark from God-consciousness that in a trice cuts away habit and the past – if we let it! – and once more tells us: '*This* is what you might be. Take this fire, remember it and strive with all your mind.' Uranus asks us to become, first of all, a stranger to our own patterns and then, having achieved this liberation (even in part), to become a stranger to the prevailing sleep of the times and the product of that sleep, the elevation of the transient and meaningless; and Uranus will give us the sense of humour, wise mockery and firmness of mind to enable us to survive the whisperings from the strange, new angel that we have become – the enigmatic whisperings of a paradoxical poetry that finds in truth the greatest of beauties.

We will turn now to Pluto.

PLUTO THE VEHICLE
Chapter Fourteen

GOODNESS
WILL

CRISIS
INCUBATION
TRANSMUTATION
RESURRECTION

A serpent is stretched out guarding the temple. Let his conqueror begin by sacrifice, then skin him, and after having removed his flesh to the very bones, make a stepping stone of it to enter the temple.

– Ancient alchemical text.

The form is from Heaven, the energy is from Hell.

– William Blake

It's such a perfect arrangement for wisdom to hide away in death. Everyone runs from death so everyone runs from wisdom, except for those who are willing to pay the price…

– Peter Kingsley

The Planet

Pluto did not begin her life as a planet. She was a satellite of Neptune until she collided with another satellite, Triton. The collision made of her the ninth planet orbiting the Sun and, as of this writing, she has her own trio of moons: Charon, Nix and Hydra.

Pluto is a small body, smaller than Mars, and her ellipse is more elongated than any of the other planets. One could say that she is an image of concentrated energy being stretched! She rotates at a slow and thorough pace, once every six of our days. As previously mentioned, the tilt of her specially created orbit allows her to pass inside Neptune for 20 years out of 248, and during this time she intensifies her purifying work.

Like Uranus and Neptune, Pluto is a mass destiny planet, particularly related to the Will of Humanity (conscious and unconscious) and, in the individual, she aids the development of the will itself which is the capacity of character, and the character's ability to achieve an aim. Therefore, it may be instructive to examine where aspects of human will were headed in the year when Pluto was sighted – 1930 – before we look at her relevance to individual development.

The first thing that strikes us about 1930 is the amount of technical science aimed at greater penetration and seeing within. 1930 saw the first use of X-rays to look at molecular structure plus the invention of a new telescope that allowed the corona to be observed. In a strange parallel development, glass was produced which was layered with a substance that blocked reflections, and this enabled the human eye to see clearly through a pane of glass for the first time. Cameras too were further developed: the photoflash and the wide-angled lens were both used for the first time in 1930.

This was the year when the great project to clean all the mosaics in the Hagia Sophia began – a beautiful symbol of Pluto's primary brief, which is to bring the purification that will enable Wisdom (Sophia) to shine!

On the lighter side, the artists seem to have picked up this 'something new' in the magnetic field of knowledge. The composer Leos Jánaček wrote his opera 'From the House of the Dead'; T.S. Eliot penned 'Ash Wednesday', William Faulkner 'As I Lay Dying'. Popular culture released films with titles like 'Hell's Angels' and 'Murder'. Even Sigmund Freud wrote a tome called 'Civilisation and its Discontents'.

Meanwhile, in laboratories around the world, scientists perspired in an urgency to analyse and to split the atom.

Humanity was driving downward and inward, looking for something, something, something…

Dreams of Incubation

In chapter 8, it was told how the Pythagoreans were part of a Mystery school of 'Apollo', and that they travelled with this school to southern Italy, where they established a powerful system of Western philosophy. However, their philosophy school was not the exercise of mind we have come to associate the subject with: theirs was an exercise in *being*.

Now, when Pythagoras visited his school in Italy, he lived in a house that possessed a purposeful basement, the function of which replicated the function of the caves found beneath the temples to Apollo all across Anatolia, Crete and around the Black Sea. For the 'Pythagoreans'[170] entered a system that initially demanded that they question every part of their lives. They stripped away personal illusions and constructs about the world and, bit by bit, began to experience the silence into which we are born and to which we all must return. The temple caves were the setting for the final process for some. They did not go there to die physically; they went into the senseless dark to wait until they died inwardly, died to every last vestige of personality and opinion.

In this process, which was overseen by a priest who had passed through this death, some contacted the true *seed* that is the human being; and in becoming the seed, matrix of everything, they were able to contact *something else*: a great truth that came to them, oft-time in

---

170 They have since become known as Pythagoreans but, as stated in chapter 7, we cannot know all the source streams feeding the flow of Wisdom at that time – and which included lines from the Eastern Shamanic traditions – all of which stem ultimately from Origin.

dreams of the transformational laws which are the bedrock of true civilisation and upon which civilisation must rest if it too is not to die in meaninglessness.

When such a person emerged from the temple, that person had become a prophet and deliverer of laws; moreover, they had *healing embedded in their words.* For that is what Prophecy is after Pluto has finished with her: it is a speech so compressed that it bursts with nourishment for ultimate human need. Such speech heals.

The whole process was called Incubation.

How superficial, by contrast, is our own desire for a true Will, its muscularity and its bravery. We in the West jump out of aeroplanes, as tourists attend sweat lodges and float tanks, walk over coals in a barely understood trance dance. We search our silent retreats, take on mountain cave experiences accompanied by weeks of meditation (years in some rare cases). What are we after? We are after that which we cannot face – the NOTHING, the tumble into dark without buffers – and so we create 'virtual' cave experiences for ourselves. It will never work. The only thing that might would be to go into the Hell of Nothing and stay there until Something Else grips the gut, if it ever does. Maybe it will. Maybe it won't...

Pluto is the hand gripping the gut.

When we have a Pluto transit to any planet, *everything* associated with that planet must be called into question, all of our fairy stories. If it isn't, if one is passive, Pluto will bring its crisis. Pluto to Mercury may bring on the madness that you have been avoiding; Pluto to Venus may let you know that you are not loved, if love be the superficial act you have been playing at.

She's tough, our wise friend Pluto. She transforms personal, wilful desires into capacity.

She is midwife to the Resurrected Will.

You see, with some will, we may be able to do some purifying Good in the world, may be able to persist until the aim is accomplished. This is why Pluto is assigned to The Good. Good is something one *does*, not a passive state; and the acquisition of Will and the Good requires many healing deaths. This is what Pluto assists us with.

The people who have powerful conjunctions or hard aspects from Pluto to personal planets are sensitively attuned to the benefits that may flow to us, if we could only re-embrace our tradition of silence and will development. The trouble is, in our intellectually sanitised and distracted world we do not have the guiding system and its master teachers, therefore, we only get compulsion and a panic about death pushed to the back of the mind. There are few who have the compressed speech to state the problem, let alone the prophetic will to instigate the necessary rectifying laws. Meanwhile, the world goes to hell.

The Transmission of Panic

One group that is particularly tuned into the collective panic and likely to enact it in their own bodies is the tribe of those with the Moon in a challenging aspect to Pluto. These are the resonators (Moon) of our collective panic. In particular, I have found that artists with these aspects bring a subtle, additional layer to their work that is somehow threatening to cultural norms represented by the critics and judges who attend the work. They can 'fail' in public[171] because the work has succeeded all too well!

Here are some examples.

Josie is a prominent choreographer, with a Moon conjunct Pluto opposite Mars. Josie is an insecure person, so transparently so that one interviewer remarked that Josie appeared not to think she would get the position (head choreographer) for which she had applied and was qualified. However, Josie says that she dips into her panic to draw out her fine creative ideas. Her early work included a rendition of 'The Firebird'

---

171 The real failure of course is in their self-perception because Moon-Pluto people carry an insecurity in a personal defensiveness, regardless of demonstrable ongoing achievement.

which is a dance about a community of men who have a tradition in which they sacrifice a firebird – only to have the bird turn upon them and survive at the end of the ballet![172] Josie and her dancer partner (a man with a Mars-Pluto square) later became committed to an extreme form of Japanese dance training, which they applied to their renditions of the darker Western classics, such as Macbeth. These productions were performed as invocations, with their dark and magical atmosphere generated by stylised silences.

William is an illustrator of children's literature. He has a tightly conjoined Moon, Jupiter and Pluto rising. William has repeatedly been passed over for prestigious listings and awards, even though he continues to work consistently. One day, he asked an awards committee member why his work had been passed over for an international children's book fair, and he was told that there was 'something about the faces of the animals, something in their expression'. It seems that his (for example) cats on the run from dogs had an oh-so-subtle additional layer that connected cat panic with that-from-which-we-flee! William has even had Saturnine judges place his work in the wrong prize category and then scream at him when he pointed this out. William has an easy, kind personality but it seems that, simply by being there, his work stirs up collective denial.

Mary is a writer and poet of rebirth. Like many Moon-Plutos, Mary has found herself repeatedly in circumstances in which others present their darkness to her and unconsciously invite her to take it on for them. Such is exhausting and Mary has removed herself physically from a place several times, without telling people where she has gone. Mary writes about the pain that opens the human heart. In one book, we find a figure inside the heart itself scarring its history like a wall being covered with graffiti. She writes of the dark wisdom expressed by those in

---

172 The subjects chosen by artists are often reflected in the challenging aspects of the chart. In this case, the firebird is the Moon-Pluto drawing on unfathomable power at the last moment in the face of the Martian pursuit.

extreme circumstances, those so far away from any hope of status that they live an emotional truth, with all its black humour…

There are many such Moon-Plutos that I have met over the years. Most do not know the gift they carry *for us* – because we don't know ourselves – and they suffer. Many project the Pluto and may become victims (Moon) of the dark that they see everywhere. I knew an old woman, who had been ruined by this, made mad. Everywhere she went, she would be drawing attention to the fearful and negative surrounds. One can't live like this' she said, 'all I want is to hear healing words, healing words, healing words…' Yes.

The Fertile Regions

The word 'origin' does not refer to historical beginnings: it means the essence of a culture, the seed. Thus, when a culture tells its creation story, its origin, it will break into its own compressed speech. This commonly takes the form of poetry, myth-imagery and song.

Pluto encompasses the mysteries of compression and the creative power of Origin; therefore Pluto rules fertility, the mysterious source of physical life.

There are many samplings of origin. Oriental creation stories tell that the world is generated in the space in which the contact between being and non-being takes place, in a world of creative chaos. One branch of Kabbalah informs us that such a space for creativity-from-chaos came about due to a contraction of Divinity: Mystery itself had a primordial experience of contraction, darkness, fertility and birth. Our oldest known creation story, the Babylonian 'Enuma elish', tells us that the human beings who arise out of this space are called, first and foremost, to do the good of the gods (to serve their Good) so that the gods might rest at ease in their hearts.

I have found that it is common for a client to give birth or become a father during a transit of Pluto. One client of mine even became a first-time father during a transit of Pluto over his tenth house Mercury. The transit upgraded his name: he became 'Daddy'. Another client, in her fifties, was due for a transit of Pluto over her seventh house Mars. She was alone and worked successfully in the corporate world, and she was hoping that this powerful energy might bring an equally powerful male partner into her life. As it turned out, her daughter-in-law gave birth and my client became a most fulfilled, first-time grandmother of a baby boy! Meanwhile, she began to research the hidden history of the men in her family as part of a doctoral thesis, spurred on by the fertilizing spark of the new baby boy and the hope of family regeneration.

It has reached the point where this astrologer will check any client's birth intention, should a Pluto be on the horizon.

Fertility, of course, applies to the soil and its regenerative powers as well. Fertile Plutonians can be gardeners and soil scientists; worm farms, compost, decay and rebirth are all of Pluto. These people seem to have the gift of bringing tired landscapes back from the dead and words like 'implant' and 'extraction' mean a lot to scientists of the soil. It is a wonderful way to work out a fertile and prominent Pluto, this renewal of sap, this mulching and weeding. There is a clue here to how we may conduct our lives, for all cycles of life decay eventually but if we remember to let go consciously, an essence of the experience of the cycle may be retained, and this will become a ferment that provokes the renewal of life. The healthy Pluto will look for the *distillation* of each stage of experience and for where wisdom may show her face at the very time of passing. This seems to be a law, this uncovering of essence right at the time of release.

The fertile Pluto is the rebirth of the world.

Pluto in Hell

Those will challenges from Pluto have a compassionate and attuned nature with a need, above all, for emotional honesty. Given this, it is sad to report on the nightmare they can become, should the regenerative potion fester. Let us look at what Pluto may become when she forgets her transformative brief.

The most obvious characteristic of a person with a stuck Pluto function is his or her paranoid notions about personal 'territory', which awareness distorts all their thinking and much of their behaviour. Even simple interactions are run through a filter of minor power trips. For instance, the person will withhold vital information or seek to separate others who have much good in common, playing 'divide and rule' or seeking to possess one party. Relationships are often formed by creating an alliance against a third person.

What about loving relationships? Negative Plutonians don't fall in love: they fall into obsession and act it out by becoming locked into what they perceive as the insufficiencies of the partner and their own pseudo-compassionate desire to 'help' by encouraging the partner talk about his or her 'problem'. Addiction to turbulence, negativity and bondage are the results, followed by acrimonious break-ups and unforgiving stories told ad infinitum to any friend who is not too exhausted by now to listen. Then Pluto becomes caught in the net of desire once more and repeats the whole scenario.

Pluto leaves disaster in her wake: she is essentially *destructive*.

When Pluto squares or opposes a personal planet, it is often the Pluto that is projected out (although not always). I have known a Sun-Pluto who has a job as a 'trouble shooter'; a Moon opposition Pluto woman once told me that she had followed a man, whom she did not know but instinctively knew had 'bad energy' (her words). Why did she follow this stranger? She wanted to 'know where he would go'.

Why would anyone follow bad energy? However, in a sense this is what all negative Plutonians do: negativity is their food and they are forever looking down into Hell. In this case, it is significant that this same woman – highly intelligent and with a spiritual direction – could never locate her true passion in life; she did not know what she really cared about and she recognised this as a problem. Pluto is the planet that puts us in touch with the voice of our own *true* passion, our Good, but we need to transmute both territoriality and projected negatives in order to hear her.

When the planet in aspect to Pluto is projected, the Pluto will attempt to divide and belittle the projected planet. One client with Venus opposite a tenth house Pluto was always drawn to people who were in the midst of expressing pleasure at seeing each other. He would then sneer at their (quote) 'superficial hail-fellow-well-met rubbish'. Another Plutonian had a habit of positioning himself as eavesdropper and punctuating the comments of others with sarcastic remarks. Mercury square rising Pluto. Everyone wants to get rid of these (frankly nasty) people; then Pluto will complain that 'no one lets me in' and withdraw into her strange combination of isolated victimhood and charismatic superiority.

Of course, when Pluto gets hold of something as vast as spirituality, she will occultise it. She will gut the Glory and retain the potential in the religious knowledge for power. Negative Pluto will not serve God: she is searching for a means of making God serve her. (And be careful if she thinks you are 'God'...).

In the story of the final battle between the Fallen Angels and the Archangels, it is told that the leader of the Fallen, Azazel, is bound by the Archangel Raphael, who is the angel set over the diseases and wounds of mankind and who brings a blessing to those who set their dependence upon spirit alone. Azazel's punishment (having led an army of rebels to earth) was to be made to take on the corruption and accumulated toxins from the community – the corruption he and his angels had brought in their fall – and then to be sent into the desert bearing it all.

Perhaps this is how Hell works. We are condemned to a desert of our own defilement. As Jesus said:

> There is nothing from without a man that entering into him can defile him: but *the things which come out of him, those are they that defile the man.*
>
> *– Mark 7, Verse 15 (emphasis added)*

Plutonians create their own Hell; however, they are also the tribe that may develop the capacity to get out of Hell. They are, after all, the custodians of the Will of the Angel Body. Should Pluto at last *name Hell as Hell* and stay there knowing she is in Hell, she will recognise at last that *character is the only territory we possess* and begin the journey upwards.

Transforming Projection

Therefore, how can a person with a problem Pluto rise?

One instance of the projected Pluto in a conscious person is the person who becomes a healer. The Plutonian healer is always an expert diagnostician, capable of naming and treating physical and psychological disease, and many fine medical and psychiatric researchers have a strong projected Pluto that is also stabilised by other aspects in the horoscope.[173]

However, if Pluto remains projected and unconscious, there will be two clear results. One is that the projector will have no practice in the art of 'dying', having always given that task to others, and may become very much afraid of physical death. The other consequence – and this is most important – is that *the projector will focus on the untransformed features of the other* and this, by projection, fault in another the very energy that they need in order to further develop their own will.

How does this work? Here are some examples.

Indian tradition tells us that there are 'six crocodiles on the river of life'. These are: pride, attachment, envy, greed, lust and anger. You could say that these are our corrupting emotions and we must all meet them and defeat them at some stage – and the crocodile is a fine symbol of a dangerous Pluto!

---

173 Dr. Carl Jung himself possessed a Yod – a Finger of Fate marking a key direction in his life – which pointed to a Fourth House (tribal roots) Pluto in Taurus from a base of Mars in Sagittarius sextiles to Jupiter in Libra, a combination with a mission for the future! Jung's Pluto was in turn square to Saturn in Aquarius, the position of the pattern of the Mind of Man, indicating his focus on penetrating to the deeper structures. I myself have a client, a twelfth house Sun opposite Pluto, who is a doctor and researcher into hidden factors behind international plagues. In this case, his Sun finds an outlet by trine to Jupiter/Neptune/Node conjunct in the Eighth.

Suppose I have a Pluto opposition and suppose I have of late begun to accuse a friend of being 'consumed by anger'. I may have had unkind thoughts about him, told myself that he is behaving like a bully. Yet, it is only in the past year that I have begun to judge him this way. Now, the common view of projection would conclude that I am the one who is angry and am recognising it in extremis in my friend. However, this is not the way *transformational* projection works.

Let us look at the word 'anger' as an energy. Isn't there something vigorous about it, something that implies a deep desire for action? If we purify the quality, what do we get? Perhaps it is the quality of *courage* that is lacking in my life? When was the last time (wuss that I am) that I confronted someone with whom I have unfinished business and finished it? How pathetic have I been lately? Perhaps it is time to write that political letter to the broadcaster, expose my opinion and take some risks? Perhaps I am oversensitive to anger of late because *its higher manifestation is the very quality I need.*

Our irritations with and judgments of another provide a road map through which we can recover a lack in our own life, if we can learn to transform and incorporate that which we judge.

When I ran a workshop in Sydney over twenty years ago, this exercise was utilised to help the group come to terms with projection. One woman present, Alison, remarked that she was forever criticising her husband for being attracted to and distracted by every superficial spiritual fashion: he was 'driving her mad'. During the process, she came to see that she herself had become such a closed-minded moraliser that anyone who displayed an unusually open mind was bound to draw her fire! She admitted that it was the openness of her husband that had first made her love him and that, through him, she had had many wonderful adventures. Alison knew that it was time now for her to try something completely new for herself (not through her husband) and get rid of her prejudice.

Most traditions recognise that there is immense power in knowing the true nature of something (or someone), which is called the power of the Name. Pluto governs such knowledge and in knowing, she can *turn pain into power.* This is how Pluto acts as a healer of the self: she names and transforms both personal, stuck territories and projected negatives.

Direct diagnosis is important; but naming an energy in the light of what we might need ourselves not only eliminates judgment but reveals again that which lives: with this new force, we feed the will.

Crisis, What Crisis?

Isn't it extraordinary? The extent to which humanity refuses to face a looming crisis? The fear of naming it and making follow-up policy, the denials or the great silences beyond lip-service are all too familiar. Well, what do we expect if we have never confronted death and haven't developed Will.

When the Mystery school that is signified by Pluto was released for examination by astrologers in 1930, two crises were just beneath the surface for which a study of the Mysteries of Pluto and the Will might prove a corrective. One was the rapid extinctions of plants and wildlife, under pressure from the new attitude of 'scientism',[174]due to fast-track to climate change.[175] The other crisis was the military application of atomic fusion, regardless of its potential to snuff out lives and bring devastation.

At that time, Pluto was tracking across the Taurus-Scorpio axis (nature and covert forces), but the potential in the dark froze because we, as always, were in denial and denial reverses living energy and causes it to stick. Our denial froze Nature herself and gave the search for source power a literal meaning. A Festival of Destruction was underway

---

174 Buddhist astrologer and scholar Patrick Curry gloomily remarks that extinction is the final product of the Man of Reason's insistence on certain knowledge and the mastery of nature through commodification, domestication and exploitation.

175 The world was warned about climate change over 50 years ago but it seems that it has taken the prolonged square of Uranus in Aries (exposure) to Pluto in Capricorn (geological climate and political territories) to enable the increase in weather events to be acknowledged and grass root organizations to challenge power. This square also represents generational change.

and neither governments nor the organised citizenry had the will to turn the ship around. At this moment, the mysteries of Pluto were revealed.

Pluto is the higher octave of Mars and calls for the transmutation of conscious and unconscious forces. What we are facing is extremes of the life force (Mars in Aries, creating burn-out) and extremes of the death urge (Mars in Scorpio as collective implosion). Together, they constitute the Shiva dance of Being and Non-Being.

Let us think about Mars. It is known to have warrior energies and to indicate an individual's method of confrontation, what kind of energy they might have available and how they will channel it. Now, the system called The Work, from G.I. Gurdjieff, has an interesting teaching about war. It says that the many deaths that occur during war release an energy that is somehow needed for regeneration. The Work is not teaching that war is good or unavoidable; the teaching submits that this same energy can be manufactured by a human effort to develop will and that it is this effort alone that will limit the need for war. To state the obvious, we need to regenerate ourselves, not project onto and fight each other! This is a most interesting teaching in light of the Higher Octave relationship of Pluto to Mars.

What then can we do? How can we co-operate with Pluto? During any transit of Pluto, we can work to eliminate the false and cure the negatives, find out about compression and conscious letting-go. Above all, we can make an effort to contract our lives around the essential (and banish substitutes with a dose of dark humour) and learn also to tell the essential truth. Right action will follow this: we will be guided.

Often one can feel the steady approach of a transit of Pluto. Our instincts tell us we are ready for change. A situation is headed for a crisis. A way of being is nearing its karmic consequences. We *know*. I had a wise client who was due to receive a Pluto transit to her afflicted fire sign Mars. This client was dimly aware that her tart belligerent rudeness had alienated her from many people and, as Pluto appeared on the horizon, she took the courage to contact all of them, state clearly her errors and apologise without excuse. She turned her sword into an effective scalpel. Another client received Pluto to Jupiter. Afterwards, she told me that she had finally eliminated her fear of physical death, her words: 'I am looking forward to death: it is the next adventure'. Looking forward to death was a major fruit of Pluto to Jupiter.

It is only through embracing our extremes as a living journey that we can hope to redeem the crises of our times. The mysteries of the Will can only be celebrated if we are prepared to enter the underground temple.

Resurrection and the Duty of the Knight

It has been said that Pluto is midwife to the resurrected will and that, if we speak to ourselves from essence, we will be guided. What is a resurrected will and how might it be guided?

The resurrected will is an expression relating to a tradition which stretches back as far as Abraham and beyond but is more recognisable to us as the Duty of the Knight. However, it is not the isolated history of males in armour seizing Jerusalem, nor modern men and women being tapped on the shoulder by monarchs about which we speak. The true knighthood is carried by both women and men who have died to their former life, come forth from the initiation of Pluto, and so become qualified to act.

They are qualified to fight on behalf of the energies that transfigure the world.[176]

In the Shiite and Ismaili wisdom tradition it is told that the great age of the prophets and teachers has come to an end and that we will have no more of them; and it is told also that, just as that great age ended, another began. This is our present great age, which is the age of *spiritual initiation* when the inner meaning of the teachings is revealed to

---

176 I am indebted to the Temenos Academy Journals of 2008 and 2009 for fine articles on the tradition of spiritual chivalry and the esoteric knighthood. See Henry Corbin, 'Youthfulness and Chivalry in Iranian Islam', Parts 1 and 2 in Temenos Academy Review, Numbers 11 and 12, Kent England.

those who have become capable of receiving such meaning. [177]

This is the age when the Divine Wisdom, the Glory that transfigures, breathes her perfume upon all our knowledge so that it comes alive. Those that receive this breathe are called the *Friends of God*. They exist as a network of people who have gone into exile from the world of Saturn and his striving after form alone. However, the exile of the Friends is not an escape or an abandonment, no. The exile is a gathering of forces in order to turn around and fight. The knight has a set role in the world and the Friends take up that role, *carrying their exile and death within*. The Friend is the being, the Knight, in action. Their weapons are a complete possession of the self, silence, compressed speech, refusal to participate in illusions and a presence that compels, by its magnetism, the devils besetting our times to reconsider their devilish lies.

The duty of the knight is to serve the Heart, the Knight's 'Lady', who is nothing other than the Glory, with whom the knights commune now as a friend face-to-face and on whose behalf they enter the drama. These do not need to consider whether something is alive or dead: they know, they carry both in their being. Their very fight has Life.

The emblems of the cultural form of spiritual chivalry began to emerge in both West and East across the Medieval period, when the knowledge was given in the form of epic stories. The great Sufi parable 'The Conference of the Birds' relates the quest to find the Celestial Self (The Golden Bird in the story), who is the angel in potential, who awaits us in the desert which is all that remains after our illusions have been shed. The story of the Knights of the Holy Grail is also a quest for the spiritual self.

How does Pluto fit into this? It is the activity of Pluto that enables a steady movement toward the moment when our own quest will be

---

177 There is an echo here of the Great Age of Aries with its individual enigmatic prophets and beginnings, and the Great Age of Pisces with its religious and symbolic social frameworks (and its confusion). The Age of Aquarius is a spirituality of Friends that is the 'as above' possibilities of the sleep-engendering technology worship of the 'so below'.

revealed to us by the Angel we have become. For the willing, Pluto will slay our dragons and transform our compulsions into character.

There is a movement in modern times called Sacred Activism, which seems to be a worthwhile effort to step down (make accessible) the possibilities of the Friends and to apply knightly activities to modern dilemmas. The movement incorporates simple suggestions about what one can *do* as an individual today to help rectify injustice and dis-function. This movement not only sets out to do good: it also calls upon the protection of the feminine face of spirit, which indicates that its activities may be of the Knighthood. As sacred activist Andrew Harvey remarks:

Someone who chooses the transformation and radical embodiment of sacred activism will not have an easy life and guaranteed success. What she will have is the certainty of being on a supreme Sacred Adventure.

Therefore the final goal of Pluto is to qualify us to fight on behalf of the energies that transfigure. How might we recognise such energies and through what guide? What else might be required, apart from Will and the awakened Uranian Mind?

Ah. We have arrived at last at Neptune. We have arrived at the station of the Heart.

NEPTUNE THE GLORY
Chapter Fifteen

BEAUTY
HEART

SEDUCTION
BETRAYAL
REDEMPTION
ANNIHILATION

In truth, there has been a great destruction of hopes in the West, and there is no telling where this will end. Its most alarming symptom is the pious agnosticism that is paralysing excellent minds and inspiring in them a panic terror before everything with the suspect aroma of 'gnosis'.

– Henry Corbin

When I use the world 'God' I am not talking about anything that can be separate from anything else.

– from Gangaji, The Diamond in Your Pocket

The Planet

Once upon a time, about 4.6 billion years ago, there was a great breath. This breath blew and blew until it blew away all the misty gases that had been until then spinning chaotically across the universe. Then as the skies cleared, the angels saw that the outer planets had been created at last! Time moved on again. The inner planets formed, the Earth, the

Moon, and water arrived on Earth. Everyone orbited together in peace. However, at about 4 million BC (Earth time), Neptune became tired, tired of struggling against the orbits of the emperor Jupiter and the lawyer Saturn. She complained and her angel moved her out, out into the Kuiper belt of icy bodies. Here she rested with her moons. Time passed again. A squabble between Neptune's eldest, Triton, and the upstart Pluto caused Pluto to leave home and to begin her own tribe. And so Neptune drifted on and Triton with her, the eldest in her round of fourteen moons, and she wondered if she would see Pluto again...

When Neptune appeared in the telescopes of Johann Galle and Urbain le Verrier in Berlin in 1846, anaesthetics and hypnotism also appeared on Earth. However, there was as well another humble happening that year. 1846 was the first year that saw Christmas cards sent along the axis of loving relationships, as tokens of goodwill.

Perhaps it was time to reconsider not-you, not-me, but the indestructible web of love in which we find our true spiritual home.

In astronomy, Neptune is set apart from the other planets by her dramatic weather patterns: ever active winds up to 2,100 kph (1,300 mph) across an atmosphere with a high proportion of 'ices', ammonia and methane, and black spot storms that come and go rapidly. She possesses an axial tilt compatible with Earth's (and with that of Mars), Neptune holding at 28.32 degrees and so seeing similar seasonal changes to that of Earth (23 degrees), relative to her frosty outer regions.

Her orbital year is 164.79 of ours.

Like Uranus, Neptune possesses a tilted magnetic field (though not as great) and a faint ring system that presents as reddish, due to the silicates and carbonates coating her ices. Neptune herself is coloured for us – a beautiful, radiant sea blue.

She is the astrologer's gateway to the Mysteries of the Heart.

Imagination or Imaginings

At the beginning of this book,[178] the reader was invited to consider the Earth as a field of Glory which is spiritually alive and whose waters, grasses and hills are addressed as beings, as 'thou'. The Mystery school that is Neptune is concerned with the space in which a dialogue between the Glorious Earth and the Angelic Self might take place. This space was called The World of the Imaginal by the mystic artist William Blake;[179] the alchemists call it Imaginatio Vera; and in inner Islamic studies (referring back to Mazdean angelology), it is known simply as the Interworld.

What is this place and how may our knowledge of it add to our knowledge of Neptune?

The Interworld straddles a region between Intellect (the world of Uranus) and Senses (the fertile world of Pluto), although she penetrates both worlds. This region is the seat of the Heart and from it come the languages of the heart: the great myths and spiritual stories, the Psalms, the parables[180], the music that heals. In the region between intellect and senses, we yet possess – although we do not know it – an Atlantean world, submerged, lost to modern education.

The Angels proceed out from here. It is the veiled abode of Feminine Grace.

The action of Neptune directs us toward this great reality, greater that anything we have hitherto taken as reality. However, our journey to such a place – really a state – must first of all encounter Uncertainty. This is *the* great problem in the field of Neptune. For along the road we will encounter the products of our own fantasies and delusions, and these will appear to be the same as reality to our undiscerning eye. One

---

178 *The Sap of the Sacrificed God* in chapter One.

179 Blake saw the seat of Jesus exactly here and referred to Him as 'Jesus the Imagination'.

180 G.I. Gurdjieff tells us that, to develop and control the emotional centre, we must visualize and not merely think and that this is why parables are visual languages.

has to become *capable* of detecting the difference between, on the one hand, our oh-so-creative imaginings and, on the other, an image-thought that houses a true Spiritual State.[181] Those who acquire the discerning eye, the capable, are those who have passed through the many pains and deaths of Pluto until the vehicle has acquired the spiritual 'muscle', whence the Resurrection provoked by Pluto causes the perception of the true things of Neptune to open; thus they are able to cleave to the Beautiful and her whispered angelic languages, right in the face of phantom temptations to do otherwise. In this struggle, it is the emanations from the Interworld, the myths and music, which are often found to be the gifts that keep our longing filled hearts steady.

How hard it is to step down Neptune. Neptune is the guest at the door that nobody wants, shape shifter to our sight, liar to our ears; and so she enters as a mist, reconstituting herself as the ghost of unwilling sacrifices, of waste, loss, poison, decay, of betrayal and of fantasy. How hard it is to step down Neptune and yet oh, how we long for her fruit.

What strange forms our longing takes.

A Forest of Fakes

Nature is a great teacher. Take fish. There is a fish in the ocean called the blue streaked cleaner wrasse or *labroides dimidiatus* (known simply as the cleaner fish) whose job it is to clean the inside of the other creatures. It does this by appearing in front of them and making movements that signal the other fish to open their mouths, whereupon the cleaner fish enters and does its work – a marine toothbrush! However, as in everything else in life, there is a fake! The ocean possesses another fish, the aspidontus, and it has learned to imitate the cleaner fish. The fake cleaner fish presents identical signals to the others in the ocean but when

---

181 Shiite studies tell us that we may ask questions of the things of the Earth in their glory and that these things may disclose to the capable a vision of their informing Angel. This is a disclosure of REALITY.

they open, the aspidontus takes a bite from the mouth and disappears.

This suitably scaly story is a parable about Neptune.

For Neptune asks us: In our own lives, how are we to tell the difference between the true and real and the imitation and how are we to select the true? Moreover – and this is most important! – how are we to recognise and banish the fake voices in our own souls? There are no set answers to these questions and of course Neptune can never express as formula. However, in our exploration of the Mysteries of the Heart, it might be helpful in this instance to examine the fact of *seduction*, for the imitation can only come to a seat at our table through this door.

The Hermetic writings tell us that the whole of the heavenly and the earthly realms are populated with both good and mischievous spirits.[182] Jesus himself taught that when we allow the negative or demonic spirits to enter into our minds and our bodies, it is because we *think that we like them*. All manner of addictions arise in this way. We accept the substitute because at the time we do not know any better. We 'like' it, even though it has no lasting value. We fail to set boundaries and habit creeps in. Finally, a state of dependency upon the unreal becomes entrenched, along with its 'demon', who refuses exorcism.[183]

There is a particular class of individuals who need to be made aware of the difference between positive and negative seduction. These are those with Neptune in hard aspect to the personal planets and, in my experience, those with Neptune to the Moon, to Venus and/or to Mars need to be particularly wary and self-honest. For this tribe possesses a

---

182 The Hermeticum teaches that the difference between the good spirits and the demons is that the latter never respond to prayer.

183 In case this sounds dramatic, a simple illustration may serve. If, say, a morning cup of coffee – or two or three – is an entrenched habit (as it was for this astrologer for many years), it is because mental alertness is sought but aggravated nervous energy is substituted. This is liked as an effect until the following morning when one awakens with 'brain fog' and another coffee is needed to clear the mind. The mind is not clear – and one day this will lead to illness. Tragically, part of the entire economy of the West operates via junk substitutes, and other economies are encouraged to put aside legitimate agricultural needs in order to 'supply' us.

gift and that gift is the ability to get inside the psyche of others and to fathom their secret longing for the sublime; then the Neptunian adjusts, chameleon-like, and enacts the longing. In the process, the Neptunian becomes lost in a forest of its own chameleon activities.

Here are some examples of the consequences of Neptune at work.

Angie has a Moon in Leo square to Neptune. She is statuesque and beautiful with long raven hair. Angie has been married three times and had several affairs. She has described to me her pleasure at looking 'drop dead gorgeous' upon entering a room and then being 'totally unavailable' – the true pedestal-identified imitation goddess! Angie's marriages were responses to seduction by flattery and each proved worse than the last, as Neptune inevitably turned and revealed the toxic face beneath the mask of apparent goddess-worship. Her final marriage saw her husband accuse her of 'madness' in court, as he sought to take custody of their child. This man has Venus square to Saturn and had the accompanying conquer-and-abandon hatred of women necessary to finally snap Angie out of her imaginings. As of this writing, Angie is alone 'by choice' and wary of approaches that begin with the man telling her that she is *special*.

In contrast, Kez's Moon in Leo square Neptune has been part of a conscious path. Kez is fully aware of her ability to seduce by enacting the anima of any man and she has been involved with a number of interesting and different men without either herself or the man taking it too seriously and with both recognising the impermanent nature of seduction. Kez finally met a good and steady Sun in Leo (conjunct her Moon) in her late forties and had a belated fairytale wedding. Kez is happy, having removed her Moon-Neptune to a home with her husband, which is located between a Leo Moon arts-filled glamourous city and the oceans of Neptune.

Sammi is an actress with a conjunction of Mercury and Venus square to Neptune. She grew up in a home full of mixed messages and emotional games and early on began to beat away a fear of her own madness. Sammi has a pattern of serial, suffocating friendships with other women who are either physically or mentally ill. Sammi's horoscope is a fine example of 'projection', for she attends very closely to the insecure and ill element in others and ignores their counterbalancing strengths: this of course has resulted in the dramatic collapses of these friendships, as in my experience projectees will know eventually that they are being diminished by the expectation of the projector and, in the case of Neptune, they will 'disappear'. Sammi has had several close friends die.

Phillip, in contrast, is a conscious Venus-Neptune square. When I first met him, he had Neptune transiting to other planets and was watching his former professional life dissolve in a round of misunderstandings. Phillip is a happily married, deeply religious man, and within the smoke of his fading former life he finally knew that he had always had a religious calling. He is now a deacon of the Anglican Church and a fine and loving counsellor, who has founded his life upon the recognition that all people 'only need to know that they are loved' in order to be healed. Incidentally, Phillip met his wife at a party, where she was dressed as a fairy and to which he had brought a plate of fluffy rosewater meringues – his specialty![184]

When I met Cyril (thirty years ago), he was a classic Mars conjunct Neptune male seducer, operating behind the slippery profession of 'Psychic Masseur' (!). His clients were all women and all were treated with an intimacy that bordered seduction, potential or actual. His technique was to profess to have spiritual contact with their 'guides', who whispered special knowledge to him during his work. He told me that he had always 'needed feminine energy'. Basically, Cyril had become a self-deluded vampire upon women, a sad state as none of his partnerships ever worked out largely because his unfaithfulness was always rationalised as a form of spiritual grace! In writing this, I wonder what may have become of him during these years of inevitable decline and loss of his handsome looks.

---

184 Astrologers will recognize the rosewater meringue as being a fine manifestation of Venus in Leo square Neptune.

Imogen too was possessed by Mars and Neptune in a seductive dance. In this case, they combined with her material Moon in Taurus to create an early life as a high-class call girl (easy money) and a constant attraction to 'unusual situations'. Like many Mars-Neptunes, she has been attracted also to some healthy outlets for this syndrome: wind surfing, float tanks and travel in exotic places; and she has used the rich experience to become a writer. Imogen became a relatively conscious Mars-Neptune but such consciousness has been hard won. There remains a paradoxical entrapment by the longing for and dissatisfaction with the eternal hunt for Shangri-la.

Neptune and seduction. Seduction and Neptune. We begin as gods on false pedestals until one day our pedestal reveals itself as the sacrificial pyre it has always been. Let us hope it is not too late when the ego starts to burn.

For in Neptune it is annihilation that we are seeking: the seduction and ruin by the sublime.[185] The poet Rumi courts God in a poetry of allurement; The Old Testament Song of Songs invites the bride into her chamber. How we long to pass into this chamber, how enticing the rush as we say 'this is it, this is it…'

IT.

The men who participated in the movement that became known as Gnosticism[186] and the men who made the religious hierarchies of the first three centuries AD felt an urging to hold onto the mystical wisdom they had glimpsed in the Christian renewal of the Way of Initiates of the Heart, but they were frightened. These men were frightened of seduction by God;[187] they knew in their hearts that this would mean annihilation, the abandonment of everything that the ego prizes, and so they did what all the frightened do in the ultimate mystical requirement…

185 The Sufis call this *Fana* or annihilation in God.

186 Not the fine gnostic perception of the Heart.

187 The transit of Uranus square Pluto in the sky since 2010 thoroughly exposed the shadowing consequences of this, as we locate the endless stream of child abuse in religious institutions controlled exclusively by men.

They began to write. They wrote mythical enactments of their fears and their writings contain the most incredible amount of rubbish, right alongside pearls of the Imagination. Neptune is forever in the border country. These writings of the fearful beat away seduction by the Feminine Face of God by beating off women. Put simply, their writing reflects a terror of seduction.[188] Gnosticism wrote of flesh as an evil and made of Earth's beauty a diabolical creation by a false god (really, how is this for a load of tripe?), while the incoming church with whom they fought created an exclusive hierarchy and ejected women from positions of authority.[189]

Few saw the sublime invitation and even fewer took the Way, and thus do specialist academics in departments of religious studies pour over the texts of the fearful still. But the Glory still finds her own.

They don't write much.

Betraying the Silent Elephant

Many religious traditions contain a moment in their storytelling or in their theology when their practitioners experience a final betrayal. This may happen after other betrayals but the last is particularly brutal. It is when the concepts through which we have made sense of life and divinity are found not to apply anymore; it is when we know we have been abandoned by God and we experience an unrelieved dark night of the soul.

---

188 It is significant that the women did not write. This may be because women are closer to the doorway that is Neptune and appear to have the planet's Invisibility built into their path. This of course has led to a general cultural belief that they were not there. Regarding the invisibility of women, Gersham Scholem has remarked, in a book on the Kabbalah, that the fact that there have been few women Kabbalists has freed this developmental path from the feminine tendency toward 'hysteria', a comment I found singularly offensive (although not offensive enough to become hysterical about it!); perhaps the women who did produce texts such as The Cloud of Unknowing (a most Neptunian title) had begun to express their own collapse into Love and its Annihilation in utterly appropriate tears? This is my surmise.

189 These hierarchies were imitations of the Roman military hierarchies of the times.

The Sufi parable 'The Conference of the Birds' sees a flock of birds pass through seven spiritual stages, each of which has to be abandoned for the next unknown state. At the last, they reach their final spiritual station. This is *bewilderment*, when they know that all their expectations have been phantoms. Jesus experienced this in the Garden of Gethsemane, when he was betrayed and knew that God had broken faith with him and left him to the rabble.

You see, along the path signposted by Neptune, *all our expectations will be betrayed.* We will love the 'wrong' person – some say it is always the wrong person! – we will lose someone, give birth to a wayward child, have a friend who abandons or slanders us in the end. The poet W.B. Yeats said that the soul itself wails as it comes to the cradle – betrayed into life.

I had a friend with a prominent Neptune, Danny. He was well into his second marriage when I met him and he had five children by his previous wife, all of whom came and went to and from his home, a fairly typical situation in today's serial relationships. Danny related to me this beautiful gem of wisdom. He said that he had been sitting in his garden one day, when he crushed the leaf of a herb between his fingers. As the scent of the garden filled the air, he realised that love is a continuum and that people come into our lives and leave only as practice for the heart. The next time my friend saw his former wife (with whom he had had angry struggles), he knew that his love for her had not ended because love does not end: it is only history that travels on.

Neptune's brief is to relocate our hearts within this continuum: this is why anything incomplete in love must be betrayed, and why Impermanence is such a lesson courtesy of the transits of Neptune. It is never the perfume of Neptune that creates the problem: it is our resistance to love that falters into 'mistakes'. Denial clogs the soul so that poisonous by-products occur – the products of our demons.

There is a certain class of person who is better able to live inside the values of Neptune than are others; however, this way of living can create problems for them in our society. Mostly, they can feel invisible and unheard and they and their point of entry into life are often betrayed by a kind of silence toward them when they come to the table. You could say that they are 'ghosted'. For they represent that which can be lost in translation: nuance, implication and the atmosphere generated by loving connectedness, through a shared spirit. Note that these realities are the ones that are easy to deny.

They are the silent elephants.

I had a friend, Linda, whose live-in partner left her after eight years. She was bewildered and sought a final 'cup of tea' with him, the purpose of which was to finish the relationship properly by acknowledging that it had occurred and musing together on what they had gained from it. The man would not do this. He had 'moved on' (as they cruelly say) and so Linda was left with a sense of unreality, as if that which she knew had been present for eight years had not in fact been, as if the silence of the man was telling her that all had been a fairytale born of some delusory need of hers to fill in inner space.

Betraying the silent elephant means just this, denying what is not overtly stated because it has become convenient to so deny it. This usually comes at the time when love whispers to us that it may annihilate all our illusions if only we would love a little more.

There are some people in the Middle East, in Asia and in Japan who regard the Western mind as 'stupid'. This appears to be because we are not trained in the art of what is not said. Our nuancing and implication skills are not honed and we also state the obvious. Neptune has much to do with the NOT, that which is filled only with subtlety and with the atmosphere generated by creating relationship. It is the white on the page of the calligrapher. Neptune is a master of the beauty inherent in the unspoken. If she speaks, it is as poet with the NOT built into the word; and of course her action draws the individual to art of all kinds, whether as practitioner or receptor.

Our friends with Mercury in aspect to Neptune are particularly astute at reading the NOT and sometimes they are marvellous strategists and manipulators of atmosphere. One Mercury-Neptune client of mine always knew when a reading was over because a change in the vibration of the room told him that the key information had been presented and the 'informing angel' was departing...

There is a couple in rural Australia who founded a winery many years ago. He has Saturn and Neptune conjunct at the Midheaven and square to his Sun; she has Mercury and Jupiter opposed to Neptune. When they brief their employees, whose job it is to sell the wine, they tell them clearly that the *way* the wine is to be sold is much more important than the volume of sales. The first duty of the marketer, as they see it, is to enhance the quality of the relationship with the potential buyer and to link, by inference, this quality to the care taken in the production of and quality of the wine itself.[190]

Then there was my friend Rosie, who was Neptune-sensitive, having it in aspect to a Venus on twenty-nine degrees of Pisces, and who had much earth in her chart. This combination meant that she took everything to heart and, in particular, was unable to negotiate deeply felt physical loss (a problem for Earth). Rosie maintained a loyal humility that always created openings in which relationships could be maintained and reinforced, even if the other person seemed to reject her or 'move on' to better pastures. After some of these experiences, Rosie's mother also died and Rosie was unable to recover from this loss because this relationship had been one of unconditional love. Finally, as an earthy Neptunian, Rosie took the obvious road to annihilation. She threw herself from a cliff onto rocks.

Her body was not found for three days.

When I consider the death of a person such as Rosie, it seems that these people vanish because they embody *in extremis* the dilemma of Neptune: betrayed by those who find something better, and the nature of their love not understood by the rest of us. Then we write astrology books and describe them as addicts and failures, as maddies or too-precious artists.[191]

---

190 Of course, most have not understood this and have offloaded their beautiful wines in a slipshod and profit-driven manner. Poor Neptune.

191 When I was told of Rosie's death, it was summed up as 'a psychotic episode'- the crowning insult.

When Neptune tries to tell us something about the love inside the NOT, we usually ignore her or tell her not to confuse the issue; when she represents the enigma of the NOT for us – usually presented as a glamour – we try to use her.[192] Frankly, it might be best that we confuse the issue a little, that we become bewildered. Best we confuse all our issues and betray all our agendas, so that we can follow our hearts into uncertainty.

The Transits of Neptune

The transits of Neptune are particularly difficult to relay to a client who is about to receive them, as (which will be understood by now) the energies associated with them are lures from the Mysteries of the Heart, with the emphasis being firmly upon 'mysteries'. Neptune, as previously said, is hard to 'step down' and to advise about.

Perhaps the first piece of knowledge that needs to be given is that Neptune encompasses a state of *Impermanence* to everything except the web of love that informs our lives. Certain things – even belief – do die at this time and I have known clients to witness actual physical deaths as the planet approaches.[193] Neptune may also become the agent of decay and illness, both of which trigger our fear of impermanence.

However, it is more common for the person to experience a provocation of the anxiety associated with possible deaths (including financial deaths such as bankruptcy). One client of mine, Taz, watched his son leap from an aeroplane, only to have his first parachute fail to open; his son free-fell before the second chute caught him in time. During the same week, Taz's dog fell from the back of his vehicle and was killed.

---

192 Ask the late Princess Diana of Wales, who died in a car accident in the 1990s while being pursued by a pack of photographers. The Princess had a Moon in Aquarius square Neptune and had lost herself as The People's Princess.

193 One client saw her partner of nineteen years suddenly taken ill and die within six weeks under a transit of Neptune to her Venus/Saturn; another lost her husband to the asbestosis that he had contracted in his youth but was only diagnosed as Neptune applied to his wife's Sun.

Taz had a transit of Neptune to his Sun-Mars conjunction in Gemini - two events, two parachutes. He reported that both incidents had a permanent effect on the quality of his love. A closer relationship with his son and greater empathy for those who are attached to animals resulted.

So expect Impermanence as a gateway to the heart.

Another method of Neptune is to undermine that which is not real *by* that which is not real! This means Neptune will infiltrate the ego, its defences and its products. Much anxiety and delusion results. Often the ego will be tempted by a 'fake cleaner fish' and end up with nothing. I had a client, a businessman, who went into association with what turned out to be a false guru, during a year of Neptune transits, and lost a fortune. He later remarked that he might as well have taken the year off and spent the money on – and the time with – his family.[194]

Therefore: undermining via enticements and fraud applied to one's own convenient deafness.

It is so hard to just sit with the anxiety of Neptune in this society that believes in 'doing something about it'. However, if one does sit with Uncertainty, all sorts of magical and symbolic gifts begin to appear, along with deeply significant, unfathomable passing connections. We enter into a world of a web of archetypes - a stranger in the street is typed as a long lost lover, or like the god Pan; a theme may recur such as the repeated, literal appearance of labyrinths, a particular book or face. Gifts come to us that are tiny symbols. Transiting Neptune to natal Venus received a gift from an artist-friend, a tree made from fine metals rooted to a rock and with leaves made from deep aqua glass, courtesy of broken vodka bottles (!); a Sun-Saturn had a canine tooth that had been lost by her beloved pet made into a necklace as talisman. This client is a 'dogged' Sun-Saturn; she had just spent transiting Neptune picking cabbages in the pouring rain.

---

194 This is the advice I had given as one option, at the start of the year; but Neptune was already present and my business client simply did not hear it or return to the recording. We are blind and deaf when the temptations of Neptune begin and we are full of our own wonderful ideas.

Of course we want to do something with all this magic: it must mean *something,* mustn't it? Our structured minds want to know what it means. Well reader, it doesn't mean anything that we could ever understand in the state we are in! Leave it alone and stay with the transit.

As the penetration of the veil increases and the archetypes impact, some will come under pressure to believe they are something that they are not. One can receive projections from the crowd, be made into a saviour, a princess or a scapegoat. We can receive the prize. There are many cases where an individual during a pass of Neptune first takes on board the flattery of the crowd, only to have it reshape as the ever-fickle mob that turns upon them and 'throws rocks'. The flattery was fraudulent and the rocks illusory but never mind.

The presence of archetype-perception will cause others to be seized with inspiration and to believe that he or she has been granted a brilliant idea, and many will waste their time pursuing phantoms. Better to stay seated in the heart and allow Her to bring the project to you, if it is real. A lot of rubbish comes from our attempts to nail Neptune down and a lot of gee-whiz inflation of notions that are too abstract. The trouble appears to be that we receive 'something' and then we try to make it happen before time, before it has had a chance to step down and trial itself into a practical form, if it has any. Some simple advice to clients is always to *set boundaries* and not to commit to anyone[195] or anything other than the values of the NOT. To this end, this is *the* time to do voluntary work, to attend to your relationship with your new baby, to give money to a struggling friend and tell him that you don't want it repaid, that the friendship has been the value. It is the time to recollect and to ponder the trail of Glory that informs the life: who has been significant, who and what has had the light behind that makes up the trail, time to listen again to the music that reminds…

People can return to one's life at the very same time that loss is

---

195   One client made the mistake of marrying a man while he had a transit of Neptune and made of her his ideal in his mind. The marriage collapsed as soon as the transit ended.

214 ~ *Astrology: history and purpose*

experienced.[196] One feels haunted by the loved who have departed. We can lose people and things that have mattered but were taken for granted, only to have to listen to our hearts at last in our grief.

We seek annihilation in love; we seek where love has left its light-filled trace. We remove the unreal by means of the unreal. Our hearts go into Exile, to the land of the Not.

Mary Magdalene: a study

The story of Mary Magdalene is a fine example of the treatment accorded to the function that is Neptune, of Neptune's processes and projection-induced reshaping over time.

Mary Magdalene was a central figure in the Christian story, an apostle of Jesus Christ, and she was privileged to witness and report his Angel Body resurrection.

Mary was born in Magdala during the time of the Roman Empire under Herod. Magdala was a fishing town and it is possible that Mary herself was one of many fishmongers there: our story has its roots in the symbol of the fish. Mary become poisoned by and possessed of 'demons' and she sought out the new spiritual healer Jesus to cure her. To understand the means through which Mary became possessed, we need to understand the nature of the town Tiberius, which was the new and artificial town proximate to Magdala. For Tiberius was the first century equivalent of the lifeless, single-purpose shopping mall of today. It was surface and cultish with, for example, extravagant baths with cheap and meaningless statues of Roman goddesses (goddess as celebrity cult) and Tiberius was a magnet for the social 'trash' of Galilee who were the only strata of Jewish society prepared to sell their dignity and live among the Romans, with their surface glitter and underlying colonizing brutality.

---

196 This happened to my winery friends (p. 150) who had a much loved elderly client die of a heart attack in front of them in the middle of a wine tasting, even as old friends from their youth were returning to their lives 'from no-where'. The couple shares a Venus on the same degree of Pisces and both had transiting Neptune conjunct this degree.

Tiberius was a dead space transmitting disease.

In chapter 9 'A Question of Witches', it was told how those who are dead inside cannot manage, let alone transform, the toxic by-products of their own activities and, much as an irresponsible industrial site will seek to relocate its own toxic waste, such will find a receptor in whoever will receive it for them. This was the position of the proximate Magdala and Mary Magdalene was a particularly receptive. By the time of her approach to Jesus, Mary had been poisoned by the town and was possessed of seven archetypal demons.[197]

Mary carried a great archetype *in potential*. This is why she received projections and waste from the collective; but the excessive nature of the energies weakened and then disturbed her legitimate spiritual possibilities, enabling more powerful demonic energies to enter. This is her story and it is a story of Neptune.

Mary came to Jesus as healer and he worked with her over a long period until her demons went and her inner space was alight with the exposure to Glory. In the process of exorcism, Mary herself learned about the methods for banishing demons and it is she who wrote down the three stories in the New Testament that deal with demons. In was Mary, too, who came via the Glory within her to possess the key gift of Vision – it was she who 'saw' the resurrection (much as many of the more literal-minded did not at first believe her) and it was Mary who possessed the master-knowledge of oil and anointing, the sacrament that links Heaven and Earth, that through healing *and pleasure,*[198] reaffirms the immediacy of the Holy Spirit.

Therefore, what became of Mary Magdalene in history?

Mary's master-gifts were to be all but excised from the canon. As a finely-tuned exemplar of Neptune and its Path, she is there but she

---

197 Seven was known as the number of completion in the ancient world. Basically, Mary had a 'full house'. Seven, of course, refers back to Babylon and to the halo of seven planets around humanity.

198 I am grateful to priest and author Bruce Chilton for this insight.

is there by inference, or rather her activities are. We had to await the uncovering of the Gnostic gospels and the somewhat strange pseudepigraphia, most of all for scholars who understood that the words in the gospels spring from an oral tradition, from a culture in which *silence communicates*. The gospels imply other levels, further information, and Mary is seated right in the middle of a filled silence.

When those who established the formal churches began imitating the Roman hierarchies (first as necessary disguise, later a status-trap for themselves), those who embodied the filled silences and the other values became suspect and none more than Mary Magdalene. By the sixth century AD, Pope Gregory remade our fishmonger as a penitent prostitute – of course! What else would the gift of anointing imply and what *is* this Vision anyway? Wasn't the Resurrection literal? Thus Christianity went off the rails, stepped down too far, and those who spoke for the NOT disappeared.

Almost.

For Mary Magdalene only went into (and largely remains in) Exile – away somewhere, not here. However, there is a subtlety in the word exile and it is a perfect Neptunian subtlety. For the motif of exile is linked to *the heart's refusal to live according to the laws of materialism and literalism*. The Neptune-ruled individual really lives in exile, even as she necessarily participates in the daily round. Sometimes, her exile is overt – she 'fails', she lives on the street as flotsam, in the hospital and the asylum, for society itself has rendered her passive, sick and deluded. However, the healthy members of the tribe of Neptune – most of them – *exile themselves* and seek the voice from the NOT because they know that their treasure is indeed located in that illusive space.

We come to Mary and her teachings there. We meet other teachers and mystics of the NOT. We meet our Mary Magdalenas, our Fatimas,[199] *our veiled and our beautiful ones* waiting for us to find them as the

199 Fatima is one figure that aspiring Muslims can choose as an axis of *devotion*. Devotion is a powerful, heart-strengthening path for Neptune.

heart becomes capable. We cast off control by our culturally constructed demons. We celebrate Mother Earth and ourselves in the pleasure of embodiment anointed by Spirit.

We extend our perception until greater mysteries begin to unfold.

The Indestructible Web

The Mystery School that is Uranus teaches us that we must remain awake to the urgency of the times and to the necessity of returning again and again to the work on (and of) the Angel Body. The Mystery School that is Pluto teaches us that we must face and undergo a series of deaths if we are to emerge from the cave with character enough to realise the gifts that constitute the true birthright of the human being. What is implied by this statement is that there is a spark within the human being that does *not* die, no matter what the pressure, and that if we can eliminate the weighty debris that acts as a camouflage, this element alone will survive.

The essential well-known teachings of Jesus are to love one's neighbour as oneself and to bless those that curse you, to forgive. These are teachings about the path that will incarnate the indestructible human spark – and this will happen across the web of relationships. As Bruce Chilton has so beautifully explained it, such actions – forgiveness, love of neighbour – not only manifest the soul: they love God at the same time. The celebration of love and the act of forgiveness make up the potential web of God and it is those who have emerged from the cave in a state of wakeful resurrection that are able to reinforce its reality.

We live in a world of misunderstanding and separation. Rarely do we really perceive what another is trying to tell us. Yet the existence of the indestructible web allows a hope that we may *connect through a shared plane of spirit*. To this end, our consolidated spiritual networks in communication are critical, those who have been drawn to the web by a longing for the Mysteries within Reality, in contrast to the apparent reality of the delusory.[200]

Knowledge of and action within the true web is the redemption that follows upon Fana, the annihilation of the unreal in the fires of God. This is the Greater Mystery of Neptune. Those who voluntarily enter into the NOT and, avoiding or swiftly releasing all counterfeits, find there the Glory, receive her breath within their resurrected hearts until the heart itself becomes indestructible.

It is time now for some recapitulation and to muse further on Mother Earth.

---

200 The community of astrologers is such a spiritual body.

THE GIFT
Chapter Sixteen

I am that supreme and fiery force that sends forth all the sparks of life.
Death hath no part in me, yet do I allot it, wherefore I am girt about with
wisdom as with wings. I am that living and fiery essence of the divine
substance that flows in the beauty of the fields. I shine in the water, I
burn in the sun and the moon and the stars.

– Hildegard of Bingen

Lesser and Greater

It is time to speak about what may occur to us through the conversations
that take place within the indestructible web. The conversations may
be actual or they may seem as a private inner dialogue with the angelic
Self; or they may be transmitted by the invisible others, our teachers
across space and time.

The initiation that is the Lesser Mysteries brings man in from a state
of off-centredness to a central position in his or her life. If we refer
to our simple cross diagram, our travels along the historical horizontal
axis end and we come to the absolute middle. In the Eastern traditions,
this position is known as the station of the True Man (or Wise May in
the Tao).[201] In this process of moving into the centre, astrology acts
as a kind of prompt-script that brings an increasing the knowledge of
'heaven' into everyday life, and each major transit is a form of refinement

---

201 In Confucianism, the stations of the exoteric grades are: The Man of Letters, The
      Learned Man and The Wise Man. This Wise Man, being at the centre, is the first stage
      of the Transcendent or vertical axis.

toward Self-realization. Particularly the conscious discernment (Uranus) between the flotsam of our fantasies and the greater Imagination (Neptune) has to be willed (Pluto). Only when one acts from the centre of one's map, like a priest in a temple and without projected planets but attending to each in its own time, can the beginning of the perception that arises from the seeding of the Angel Body may begin.

As the Lesser Mysteries is our time of 'know thyself' and of course the basic astrology chart is an invaluable tool helping us toward what will be a transition. When the time of transition begins – and this may take a long time indeed and is commonly the time of greatest fear, slipping back and abandonment of the way[202] - we are asked to flex the muscle of any will we have developed and (in practice, in reality) to reverse our former priorities until we *know*. For any of us can speak (or write) well; but how do we live our lives? What makes us laugh? What do we do when we come under pressure from the mob? What are our activities contributing to the celebration of our human relationship to sacred earth? Or to the relationship of man *to his own soul*?

What are we?

Threshold Moments

The threshold to the transcendent realm of the Indestructible Web is a locus in space-time, when the progress of the initiate is apparently held in limbo while he or she comes to know the reality of the spiritual reversals by putting them into practice in his or her inner life and as action. One is required to live one's life according to and to prioritise *non-duality, relationship, necessary sacrifice* and *living in uncertainty*. These are bridging practices, not ends in themselves and so they are

---

202 One Fourth Way school teacher told members of his group in Australia that he had wanted to commit suicide when he was, as he put it 'having his turn-around'. Fortunately, he did commit suicide and became a much loved teacher instead of himself.

relatively easy to speak and to write about.[203]

Take Non-Duality. Duality - good and evil, God and devil, Cartesian mind and body[204] - produces in us and reinforces in us the function of judgment: we are always passing judgment. God is better than the Devil, mind is better than body. Therefore, a primary practice for the initiate at the threshold is to become aware of and then to arrest his or her ongoing moralizing. Morality only arises when we add time to spiritual law. In astrological terms, one could surmise that, as Chiron passes inside the orbit of Saturn carrying the spiritual gifts from the outer planets and has to turn to pass into time, it is morality that is produced by the distortion. At the level of Mystery, morality is only a wound.

A breakthrough moment can occur for the initiate when he or she recognises that *every* event is his or her life is part of The Good; it simply happened, with no polarizing label at all and, if it is received as such, will become part of a living thread of The Good that is one's life properly recalled. Again, easy to write about...but if you lover leaves you and you are fired from your job in the same week your mortgage is secured...? If one is at the threshold, the tendency is to immediately judge this as a bad week and retreat into polarity; but if there is sufficient muscle, the events may be received as that-which-happened and a vertical moment of grace experienced. The grace is the reality.

---

203 This has resonance with the experience of my Uranian friend Ben (p. 130), who knew all the Buddhist meditation theory about mandala construction but did not know it was real until he found that it saved his life because he had consistently practised it without any need for instant results or proof in advance. The demand for instant proof before entering into practice is a western disease, along with expecting to be able to argue one's way into being right about an unpractised system. Thus do heliocentric scientists argue the irrationality of geocentric (traditional) astrologers, without any study let alone practice of astrology as a Way.

204 Traditional thinker René Guénon has written: 'The Cartesian dualism of spirit and body', which has managed to infiltrate all of modern Western thought has no basis whatever in reality...On the other hand, the triple division into spirit, soul and body has been unanimously accepted by all the traditional doctrines of the West, both in antiquity and in the Middle Ages.' See Guenon, 'The Great Triad', Munshiram Manoharlal Publishers, 1994.

Sometimes, communication between threshold initiates and others can take a frustrating, if humorous, turn.

FRIEND:	I notice you have closed your business.
INITIATE:	Yes.
FRIEND:	Didn't you get enough customers?
INITIATE:	I decided to do something else.
FRIEND:	Times are hard economically.
INITIATE	WAITS
FRIEND:	It must have been a lot of work.
INITIATE:	CONTINUING TO REMOVE CONVERSATION FROM GOOD-BAD
	I decided to do something else.
FRIEND:	It's a pity. I remember you telling me about putting up the website.
INITIATE:	Yes. I have changed the site now.
FRIEND:	That must have been expensive.
INITIATE:	WAITS FOR FRIEND TO ASK ABOUT HIS PRESENT ACTIVITIES.
FRIEND:	HAS RUN OUT OF NEGATIVE SUGGESTIONS AND IS BEGINNING TO FEEL UNCOMFORTABLE.
	Well, I need to have lunch. I'd better go.
INTIATE:	There is a lovely café up the road. Enjoy.
FRIEND:	LATER, TO ANOTHER FRIEND
	I don't know what's the matter with him: he's so hard to get along with these days!

In short, polarity judgments are the food of common society. Expect a period of alienation when you abandon this system.

Then there is the threshold requirement to become witness to one's life, to become free of judgment regarding the *appearance* of the life. We know that we can fail socially at the same moment that we rise internally. In this way we are able to greet others without judging them and with compassion. These are all threshold struggles; the fruit is an increased inner stability and a tougher vehicle.

The importance of *Sacrifice* on the path has been stated. The threshold will require authentic sacrifices, those that bring one into *Uncertainty*, as no result is guaranteed and no proof can be presented to one in advance by another: the journey is yours alone and what may be sacrifice to you is nothing to another person. Money cannot be asked of those with no interest in profit.[205] Often an entire philosophy has to be sacrificed, one's politics abandoned. When your own errors and partial truths are required of you, take time out to grieve for them because the loss of a former prop for meaning in one's life is a real loss.[206]

The ascendency of *Relationship* is perhaps the most critical of threshold practices. In ordinary life, one does of course cultivate certain relationships for certain purposes, mostly for emotional security; however, the maintaining of connectedness for its own sake is the true practice, to meet with others and simply be present to what arises. The question 'will this increase or diminish relationship?' begin to come to the fore when moments of decision present themselves. In the end, we seek to expand the scope of our relationship to divinity by preserving the relationship with the divine spark in the other. One takes to the time to buy vegetables from the local growers rather than order online from the convenient mega-supermarkets, or one's holiday film may be presented to the developer with conversation in preference to our latest updated computer program; perhaps one does not uproot permanently from all positive historical connections to move to another state for the sole purpose of making more money than one needs. What are we searching for?[207]

---

205 However, it 'profits' one nothing to give away gifts from God. This is not a sacrifice. It is a waste of a blessing.

206 And yes, this will come to include astrology.

207 Ben told me that one attachment he appears to have lost is his 'attachment to location'.

The little examples here are attempts to step down into common experience what is really a difficult process without supporting formulas. The threshold is a period of Uncertainty, and it is a period when even time itself has to be remade as a Quality and as a State. One sacrifices the idea of time as being a sort of vacuum to be instantly filled in by worthwhile activities (no doubt in case we are left with enough time on our hands to bring about consideration of our own death!). It is a time when Uncertainty becomes a positive quality in present moment, not a reaction and it is the embrace of the quality of Uncertainty that is key, the embrace of the uncertain when it seeks us out.

Here is a story from the threshold.

The Doorway of Handicap

Tania is a friend who had been on a spiritual journey for many years and had done serious work on her character when we met. She was a borderline workaholic because, although her work had great integrity and had built into it time for reflection and relationship, she never took a day away from it. Tania never made space for a break in her routine. Then one week circumstances and cancellations conspired to give her a day off and Tania decided to spend a Sunday hopping on and off buses around the harbour 'just to enjoy life and have a look around'. Well, of course a break in entrenched habit is one of the best things one can do to open a doorway if one is ready – and Tania's angel was waiting for her!

This is what happened.

As Tania stood at a bus stop between trips, she was approached by a man in a loose, unkempt bus stop attendant's uniform. He was carrying a sheaf of timetables. The man had an obvious intellectual handicap, which manifested as him constantly repeating himself and stuttering and as he approached each person in the queue, he would stab at the timetables and demand to know where the person was going so that he could consult the bus times.

Now Tania, who has never had much patience, was carrying under her arm her current reading matter, which was a mystical treatise on Russian icons and, as the faux attendant stood in front of her with his timetables, his eye fell upon the book.

'What bus? What bus?' He stabbed at his papers and then indicated Tania's book. 'My uncle got those,' he pointed up the road. 'My uncle new, that book, them.' Then back to work: 'what bus? What bus?' and again pointing up the road 'my uncle'. As he continued along these lines, with repeated focus on her book and the icon on the cover and his immigrant uncle's implied collection, Tania was becoming uncomfortable in front of the other passengers and desperate to be rid of the 'pest'.

'Yes, yes, I know!' she all but shouted. 'They're icons and I don't know your uncle!' She then noticed her bus pulling away – she had missed it due to the distraction and another would not come for an hour. Inwardly blaming the attendant and judging her own embarrassment as ruining her one day off, Tania stalked away to a nearby café and ordered a coffee. There she relaxed, had a marvellous moment of witnessing herself and began to laugh.

Then the miracle happened.

As the coffee arrived at the table and the aroma reached her – and with Tania laughing at herself and her impatience – suddenly, she *saw*. Tania ran from the café over to the bus stop to find the handicapped angel who guarded the treasure, whose uncle she now knew – knew in a glimpse through the Other Doorway - was a Russian émigré mystic and collector of icons...

The bus attendant was gone, no-one knew where. Tania never saw him again.

Tania has since told me that her impatience and self-consciousness have always been her failings, her blind spots, and that the bus attendant's stuck-record manner provoked the signature of her own sticking place. It seems that the attendant was set to clear away Tania's resistance and lead her over the threshold. Tania thought that she had failed the test but later she integrated the deeper lesson. It appears that at the threshold we are tested and the test often takes place right at the centre of our own failing.

These days, Tania teaches that anyone on a spiritual path must get to know his or her central blind spot or handicap *as a gift*. She says that the blind spot in space is the equivalent of uncertainty in time; moreover, it is our handicap that is the key to prayer itself, for we pray ever through our weakness, our unformed, our inability and our miracles find us there. They press upon our point of vulnerability and if we know ourselves enough to know that this is where we 'fail', our miracles will find us out. This is a lesson of the NOT.

Tania takes most of her time off now; strangely, she gets a lot more done than she used to.

The Perfumed Land

The land of the Greater Mysteries is a perfumed land and the Mysteries themselves like a grace that is always present, even in circumstances that we believe divide us. Forgiveness graces sin even as sin apparently occurs – forgiveness is built into our failings. Love is built into enmity.[208] It is not an injunction to love your enemy that is set in biblical story: it is a call to refine one's sight so that one sees that a *good enemy is love in action*. As has been said, it is important not to confuse moral instruction with simple statement of spiritual law. Statements of law are always compressed and the things that, on one level, are thought to be separate and in need of moral control are really united in the land of the Mysteries. Love abides as continuum. It does not matter that we no longer see the former beloved or the old friend or the enemy: they are still there, The Lover, The Friend, The Enemy are all practices in love and the freedom that both informs and emanates from it.

In the chapter on Pluto, it was remarked that many who come forth from the cave of death speak little and with compressed speech. How

---

208 In astrology, the seventh house describes both open enemies and marriage partners; and the Twelfth House is a house of covert enemies and is also a house of mystical realisation.

can it be otherwise if morality has collapsed and the unity of formerly separate things – forgiveness and sin, love and enmity – has been experienced in a place where time has died?

When Jesus said to those he taught, 'I am the way, the truth and the Life', he was speaking in the compressed speech of the third level of the Angel Body – he certainly was not referring to himself, to Mr. Jesus of Nazareth in this physical body when he said 'I'. His I is an I-ness from the Greater Mysteries speaking through him, as both a presence now and a potential for all humankind. Of course, those listening thought he meant him-self (small 's'), so they nailed him up for sedition and then made a religious cult out of him. Who can blame them? Jesus didn't; he forgave those who did not know what they were about.

What follows from the coming of compressed speech is that a period of silence inevitably attends the entry into the Perfumed Land, for who on earth is going to understand what is being said? The withdrawal into compression brings one closer to the Heart, like a journey from many petals into the centre of a flower, and compression also leads the candidate to live in a realm where he or she has become not only silent but strangely invisible to the crowd. Not always – there are those who have particular tasks – but a period of Invisibility as well as silence is a fruit of a new residence in the Glory and Her State of Giftedness.

As the Buddhists have advised, kill the Buddha if you meet him on the road. The Buddha as external teacher, as polarity, does not exist.

Giftedness

In this little book, I have used the term God a number of times and pause now to point out a failing of that part of the world that speaks in a European language. In English and related languages, 'God' has a reductionist and carry-all meaning: it can range from God Most High through to a mystical state. God has even been used synonymously with Heaven in translations into Eastern languages. Yet in Eastern traditions, we find many and more subtle terms; particularly a distinction between the totally unknowable Principle and any idea of God is emphasised and so the duality between God and Man is abolished.

At this point, (and honouring the unknowability of First Principle and the reductionism of God), let us put aside God and our translating difficulties and come back to that which we can know something about – the living Earth, in the light of science – and see where this may take us in our study of spiritual astrology.

We will begin with the post-Copernican centuries, just past.

After it became accepted right across the globe that our Earth revolved around a central Sun and that our home was a part of a solar system, there was a gradual *shrinking away from the notion that humankind is somehow special*. The new narrative proclaimed that such a universe could not have been created for us as we were demonstrably not 'central'. This new narrative was not altogether a bad thing, as it robbed the spiritually ambitious of their physical justification. However, over time and increasing discoveries, another rule about man's place in the universe replaced the old religious ladder. The new scientists affirmed a Principle of Mediocrity (a.k.a. The Copernican Principle) which, as the name suggests, set man as a random spec in a sea of galaxies where (it follows) many more such creatures are bound to exist, given the same 'Goldilocks' conditions. Man had emerged as mediocre and insignificant. This perspective was reinforced as telescopes became larger and the vastness-to-infinity of the sky and its trillions of suns was made apparent to us.

The Principle of Mediocrity was rock-solid (at least in public) by the mid twentieth century. However, as has been remarked, God loves clues and about this time the very technology being used now to search the skies sent home a suggestive and undeniable image. In February 1990 – Valentine's Day to be precise – the Voyager mission beamed a photograph of our Earth back to us from four billion miles away. The image showed Earth as a distant spec, yes, but *set inside a shaft of light*, as if we were being lit by a rose special in an enormous theatre. Of course, the most hard-nosed of the public scientists described the beam as 'only optics' and 'coincidence' and reaffirmed our spec-ness rather than our specialness.

Now, it seems that it takes quite a lot of 'coincidences' to create Mother Earth regardless of her shaft of light. It takes an enormous amount of finely-tuned factors to create a habitable planet and the probability against our even existing is staggering. Let us stop and examine some of these factors.[209]

In the chapter 'The Other Science', the position of Jupiter and its role as comet-catcher and protector of Earth was noted. Without such protection, Earth would be a dangerous habitat. Therefore any other system would need also suitable giants at their border: such would be a condition for life.

There is location, which must have two contexts to allow nature to live. The first of ours is our position in the Milky Way, with its huge spiral arms and its Sagittarius galactic centre with its black hole. Here, we are discovered to be in a relatively small band known as The Galactic Habitable Zone, a safe area between the spiral arms and well away from the Galactic Centre. The effect of being in the open-country region of the Milky Way is that there is no dust over us: we are in the perfect spot to observe the infinite sky and its stars. Second, within our own little solar system itself Earth is in the only band that could support life. Five degrees further toward the Sun and we would fry, five degrees to Mars and we would freeze! And speaking of observing the skies, only Earth's atmosphere is transparent: the other planets have haze overhead. Again, *only the creatures on Earth have a clear view of the stars.*

Then there is our Moon. A planet supporting life as we know it would need to have one moon whose gravitational pull would maintain its earth at an approximate 23 degree tilt. This tilt, combined with the Earth's revolution around the Sun gives the seasons and therefore our

209 For an in-depth look at all such factors, see Jay Richards and Guillermo Gonzalez, 'The Privileged Planet – how our place in the cosmos is designed for discovery', Regnery Washington D.C. 2004.

cycles of growth and rest. The Sun too must be of an appropriate size to keep us rotating daily and not draw us inward, where our rotation would become locked. In fact, our Sun is 400 times larger than our Moon *and* 400 times further away. This extraordinary feature – one has to laugh when the materialist describe this too as coincidence – means that we are able to have a precise eclipse of the Sun, during which short time (51 seconds) we are able to observe the Sun's corona and light spectrum and so to *know* how the other suns in our vast neighbourhood work – the eclipse invites us to knowledge and our clear skies allow us to pursue it! We are fed possibilities of speculation and wonder.

It seems that only we can view the skies and know what we are seeing. Let's leave off 'why?' for a bit, as this unique property is only the tip of Earth's iceberg.

There are great forces holding our universe together: electromagnetism, nuclear, gravity. They must exist *and* they must be in balance: change one iota of any of them and we have no life! Waves along the electromagnetic spectrum, such as microwaves and gamma waves, are constantly circulating and impacting us but only the narrowest of bands in the entire spectrum is of any use to us, a measuring of one trillion trillionth of the whole.

Guess which type of band waves our Sun produces?

Then there is the physical make-up of Mother Earth. For complex life to exist, we must have: water and the correct ratio of water to land mass; the correct planetary size for the movement of liquid iron in the interior to enable magnetism; an atmosphere that is 78% nitrogen, 21% oxygen and 1% carbon; moving tectonic plates to recycle the elements; a crust of perfect size to work with the plates to promote growth…

There is more but this is not a book about geology and astrophysics.

In the 1990s, in the context of speculation about the purpose (if any) of life on Earth, philosopher Jay Richards and astro-biologist Guillermo Gonzalez produced a conservative calculation of the probability for the existence of Earth and her life forms (conservative meaning that in reality it is far *less* likely that we could exist than is indicated by their numbers). Richards and Gonzalez took each factor critical for life into account and gave it a probability rating. Then they put all the factors together. They found that the likelihood of the existence of Earth is – wait for it – one in one trillion billion, that is 1/1,000,000,000,000,000,000.

In the 1960s, the programmatic search for other life in the universe began and radar telescopes have been sweeping the skies and listening in ever since. We haven't found anything. Should that surprise us?

There is one more astonishing thing about the human being in this context, and its probability cannot be calculated. It appears that there is a continuum linking human understanding and the organization of the universe, including our own planet.[210] Put another way, it may be that the universe is created in such a way that we are able to understand it, that only we can witness wakefully to its beauty and its truth. My dog cannot know the universe in this way but I am capable of doing so. My mind has infinite reach across principles. My mind can uncover the pattern beneath the periodic kingdom in chemistry.[211]

Now, there is a consistency between this remarkable observation about the connectedness of the human being with the cosmos and the parallel astrological understanding of the continuum that exists across time-space between the soul of the individual and its *spiritual* setting in the solar system. Astrologers know that we are the solar system: we begin with this fact and then we strive, our study being celestial physics and its prophetic language, much as astrophysicists communicate in the elegance of higher mathematics.

If the probability of our existence is so slight and if the human being is constituted in such a way as to invite understanding of and participation in this rare place, surely we must recognise – and with astonishment – that our lives are a *gift*.[212] However, here we must be cautious,

---

210 The laws of physics are consistent everywhere.

211 Richards and Gonzalez point out that much of science is not needed for survival. This fact alone must bring human witnessing and discovery into play, if we are talking 'purpose'.

212 When Fourth Way teacher G. I. Gurdjieff asked the boy Fritz Peters what life was,

for a gift implies a giver and so we may again fall into dualistic thinking and the distancing of Spirit. Therefore consider: if Earth is a gift and if there is a continuum between the human being and the Way of the Universe, perhaps this tells us that our natural *state* is a State of Giftedness. Achieve this and then we will see what obtains.[213]

The State of Giftedness is our home.

It is time we came home.

Peters replied that life is 'a gift'. At this moment, Mr. Gurdjieff let out a sigh and agreed to the boy's request for private instruction.

213 Actionless Action arises from this state.

IN TRUST
Chapter Seventeen

Jesus said to them, 'When you make the two into one, and when you make the inner like the outer and the outer like the inner, and the upper like the lower, and when you make male and female into a single one, so that the male will not be male nor the female be female, when you make eyes in place of an eye, a hand in place of a hand, a foot in place of a foot, an image in place of an image, then you will enter the kingdom.'

– The Gospel of Thomas: 22

We have come a long way in our study of astrology. It is time now to do some tightening and to make some practical suggestions.

We hold a Way of Knowledge in trust and *we must believe what we know*. True belief is action in the world – muscular and persistent. It is vital that astrologers with an ongoing practice deliver the wonderful insights, which are the fruit of our Way, beyond the limits of astrological readings and into the wider community. The matter is urgent. The way of mind that we hold is succinct, prophetic and embedded. Cherish it. The astrological mind, once firmly established, will find its way into every part of the life – and this without referencing the planets at all! It will penetrate one's art, one's approach to scholarship and one's activities.

It has been told that the astrological map is a potential temple for the evolving Angel Body. Now, there are some compelling examples of other such symbolic temples in sacred history. One tradition describes such a temple, which sits at the border of 'China' (in mythical space). The temple has seven gates and a seven-tiered dome and a well inside its precinct with a heptagonal opening. The temple rises from a high, mountainous rock; and spears thrown at the Temple only turn back upon the thrower!

What is beautiful about the lore of this temple is the description of the state of those who first come before it. Henry Corbin writes:[214]

Anyone who beholds the Temple, the dome and the well is seized with a violent emotion in which impatience, sadness and an attraction that captivates the heart mingle with **a fear lest something of this Temple may be destroyed or ruined.**

Some years ago, I happened to be at the counter of a metaphysical bookshop in Sydney, when a young woman also there enquired about the astrology section. It was clear that this girl was only just becoming interested. I turned to her and, in a rush of desperation (which I very much hoped did not show) poured out my encouragement, telling her that unexpected treasures and wonderful opportunities for growth and contentment were concealed in the study, if she would spend some time with it...

We are all concerned that the Temple will not fall into ruin by neglect.

Here are some practical suggestions for Temple reinforcement and polishing, in our down-to-earth everyday lives.

Trust your judgment. When a subject 'lights up' for you, pursue it no matter how irrelevant it seems.

Reinforce networks across time and space. Find the authors you respect and be discerning in your online associations.

Pursue those who embarrass you: they are outside your range and carry spirit for you.

Never deny the Silent Elephant.

Stay awake to gifts, to receiving and to giving. Run a true economy of Giftedness.

Maintain a cycle of projects that are just a little bit too hard for you. Such will extend your range permanently.

---

214 Henry Corbin: 'Temple and Contemplation' Kegan Paul International 1986.

Keep a clear aim and friends beside you who understand it.

Clean up 'diabolical relationships'. Resist forming these alliances, which are based upon a mutual dislike of a third party. (The practice of diabolical relationships results later in the group need for a scapegoat and, later still, in mob action. Much of world politics is based upon these alliances.)

Don't waste your time on the walking dead and beware the hungry ghosts. Avoid bureaucracies.

Develop one-liners (using your planetary insight) to mess with the minds of entrenched authority. A favourite of mine is: 'Life is not about economic growth, it is about a growing experience of what we already have'. I find this confuses the greedy.

Teach the basic language of astrology where and when you can. For example, if you have a friend who teaches at the local school, enter the school as a guest. Teenagers in particular love our language.

Then:

When Uranus appears - break out of yourself.

When Neptune appears – surrender to love.

When Pluto appears – act, on behalf.

Continue these practices until they are so integrated that one does not need astrology to assist one to remember them.

At the start of this book, I spoke about the Sabians of the East and provided a few examples of the possible root meanings of the term 'sabian', (Chapter 2, note 35).

There is another possible source meaning of the word.

Nearly five centuries ago, a book appeared that claimed that the term sabian came down to us via the Hebrew word saba and that saba translated into Arabic is *sabi-un*. Now, saba simply means 'army' but saba-un does not refer to any earthly standing army: sabi-un means *the army of heaven*.

The Sabians may have been (and so *remain*) one spiritual feed in the ongoing struggle to keep open sacred space, its eye and its language, our clue being that the Sabians were so slippery to history. They left everyone guessing.

However, an invitation to cosmic battle seems a heady requirement for a humble little astrology practice. Perhaps we wish to be left in peace?

Having a Mind that sets the human soul on a continuum with the Cosmos and which translates this continuum into physical man set contiguous with an Earth that is experienced as Sacred Gift, is not a trivial possession. It *matters*. Having a practice that invites the soul to obtain to a condition where one is able to penetrate to the heart of a problem and elicit a reversal of values is not trivial. It *matters*. It's breathtaking! *Of course* this knowledge makes of one a link in a celestial army. The practice of the astrologer's mind in public space (not the astrology *per se*) is the battle and, believe it or not, once you resolve to take this path, life becomes a lot more interesting!

We are like little 'Noah-s' protecting the creatures while waiting for things to change. We are like the True Man who, arriving at the middle of the cross, suddenly sees that humanity may unite Heaven and Earth. We witness for the Earth *because* we witness for the Sky and our witnessing is as the oil of the anointer: our witnessing is unction. We witness because an indifferent Cosmos can only translate into usery on Earth. Everything is for sale. If this seems to you exaggerated myth, it may be because this text has sought to counter the internalised trivialization of our art that has accompanied its practice. The restoration of a dignified Astrology may confuse some.

Again, it is a question of learning to believe in what you know.

Then, what can we say in closing?

Continue your practice; continue to do your astrology readings until the mind steadies and a glimpse of Angel Body possibilities opens to you. Remember to care about your clients and prepare well for your sessions.

Above all, know that the astrology chart is a Song of Praise: an iconography of angel-planets written on time and blessed.

Jane Ahlquist – 2015

BIBLIOGRAPHY

Armstrong, Karen: 'A History of God', Mandarin Paperbacks 1995.

Atkins, Peter: 'The Periodic Kingdom – a journey into the land of the chemical elements', Weidenfeld & Nicolson 1995.

Austin, R.W.J.: 'The Sophianic Feminine in the Work of Ibn 'Arabi and Rumi' in Lewisohn, Leonard (ed.) 'The Heritage of Sufism' V2, Oneworld 1999.

Baker, Rob: 'Maps of Presence' in PARABOLA, Volume XV, Number 1, 1990.

Barker, Margaret: 'The Great Angel – a study of Israel's second god', SPCK London 1992.

Barker, Margaret: 'The Lost Prophet – the Book of Enoch and its influence on Christianity', SPCK/Albingdon Press, 1988.

Barker, Margaret: 'The Older Testament', SPCK 1997.

Bennett, John: 'Witness – the autobiography of John Bennett', Claymont Communications USA 1983.

Bobrick, Benson: 'The Fated Sky', Simon and Schuster 2005.

Burckhardt, Titus: 'Mystical Astrology According to Ibn 'Arabi', Beshara Publications England, 1997.

Carter, Charles E.O.: 'Astrological Aspects', L.N. Fowler and Co. Ltd. 1971.

Charles, R.H. (translator): 'The Book of Enoch', SPCK 1997

Chevalier, Jacques M.: 'A Postmodern Revelation - signs of astrology and the apocalypse', University of Toronto Press, 1997.

Chilton, Bruce: 'Mary Magdalene – a biography', Doubleday 2005.

Corbin, Henry: 'Alone with the Alone – creative imagination in the Sufism of Ibn 'Arabi', Bollingen Series XC1 1997.

Corbin, Henry: 'Temple and Contemplation', translated by Philip Sherrard, Kegan Paul International 1986.

Corbin, Henry: 'The Man of Light in Iranian Sufism', Nancy Pearson (trans.), Omega Publications 1994.

Corbin, Henry: 'Youthfulness and Chivalry in Iranian Islam', Parts 1 and 2 in the Temenos Academy Review, Numbers 11 and 12, Kent England 2008-2009.

Critchlow, Keith: 'Time Stands Still - new light on megalithic science', Floris Books 2007.

Curry, Patrick: 'Enchantment and Modernity', PAN: Philosophy, Activism, Nature no. 9 2012.

de Santillana, Giorgio, and Von Dechend, Hertha: 'Hamlet's Mill – an essay investigating the origins of human knowledge and its transmission through myth', Nonpareil Books 1969.

Field, Reshad; 'Going Home – the journey of a travelling man', Element Books Limited 1996.

Field, Reshad: 'The Invisible Way – a journey through the world of Sufi teaching', Harper and Row, San Francisco 1979.

Gaunt, Bonnie: 'Beginnings, the sacred design', Bonnie Gaunt, Michigan U.S.A. 1995.

Godwin, Joscelyn (Ed.): 'Cosmic Music – musical keys to the interpretation of reality', Inner Traditions 1989.

Gold, E.J.: 'Visions in Stone - journey to the source of hidden knowledge', Gateways/IDHHB Inc., Nevada 1989.

Gonzalez, G., and Richards, J.W.: 'The Privileged Planet – how our place in the cosmos is designed for discovery', Regnery Washington D.C. 2004.

Green, Tamara: 'City of the Moon God - religious traditions in Harran', E. J. Brill, Leiden Holland 1992.

Greer, John Michael: 'Introduction to Poimandres, the Shepherd of Men', internet http://hermetic.com/texts/hermetical/hermes1.html, January 2013.

Gribbin, John: 'In Search of the Big Bang – the life and death of the universe', Penguin Books 1998.

Guénon, René: 'Insights into Christian Esotericism', translated by Henry C. Fohr, Sophis Perennic, Ghent NY 1943.

Guénon, René: 'The Great Triad', Munshiram Manoharlal Publishers 1994.

Guénon, René: 'The Lord of the World', Coombe Springs Press 1983.

Halevi, Z'ev Ben Shimon: 'Kabbalah and Astrology', Kabbalah Society 2009.

Harrington, Daniel J.: 'Wisdom Texts from Qumran', Routledge London 1996.

Harvey, Andrew: 'The Hope – a guide to sacred activism', Hay House Inc., United States 2009.

Hidveghy, Agnes, and Oksanen, Reijo: 'Gurdjieff and Astrology', Proceedings of the 10th International Humanities Conference, All and Everything 2008.

Holloway, Richard: 'Doubts and Loves – what is left of Christianity', Canongate Books Ltd Edinburgh 1991.

Jaffe, Bernard: 'Crucibles: the Story of Chemistry – from ancient alchemy to nuclear fission', Dover Publications 1976.

Johnson, Robert A.: 'He – understanding male psychology', Mills House, Berkeley California 1989.

Joseph, George Gheverghese: 'The Crest of the Peacock - non-European roots of mathematics', Penguin Books, 1991.

Kingsley, Peter: 'A Story Waiting to Pierce You – Mongolia, Tibet and the destiny of the western world', The Golden Sufi Centre 2011.

Kingsley, Peter: 'In the Dark Places of Wisdom', Golden Sufi Centre California 2010.

Klein, Morris: 'Mathematics in Western Culture', Oxford University Press 1953.

Knight, Christopher, and Lomas, Robert: 'Uriel's Machine - the ancient origins of science', Arrow Books 2000.

Kurrels, Jan: 'Astrology for the Age of Aquarius', Parkgate Books 1997.

Maclagan, David: 'Creation Myths – man's introduction to the world', Thames and Hudson 1979.

McAulay, Anne: 'Apollo: the Pythagorean Definition of God' in 'Homage to Pythagoras', Lindisfarne Association 1982.

McCluskey, Stephen C.: 'Astronomies and Cultures in Early Medieval Europe', Cambridge University Press 1998.

Miller, Jerome A.: 'The Way of Suffering: a Reasoning of the Heart', Second Opinion 17 1992.

Milne, Joseph: 'Metaphysics and the Cosmic Order', Temenos Academy Papers 2008.

Murchie, Guy: 'Music of the Spheres – the material universe from atom to quasar', Volume One 'The Macrocosm', Dover Publications 1967.

Needleman, Jacob (ed.): 'The Sword of Gnosis – metaphysics, cosmology, tradition, symbolism', Penguin Books Maryland USA 1974.

Ngakpa Chögyam: 'Wearing the Body of Visions', Aro Books Inc. 1995.

Nicolescu, Basarab: 'Science, Meaning & Evolution – the cosmology of Jacob Boehme', Parabola Books 1991.

Norris, Ray and Cilla: 'Emu Dreaming – and introduction to Australian Aboriginal astronomy', Emu Dreaming Sydney 2009.

Ouspensky, P.D.: 'in Search of the Miraculous – fragments of an unknown teaching', Harvest/HBJ 1977.

Pagels, Elaine: 'Beyond Belief – the secret gospel of Thomas', Vintage Books, 2004.

Pagels, Elaine: 'The Gnostic Gospels', Vintage Books 1981.

Park, David: 'The Fire Within the Eye – a historical essay on the nature and meaning of light', Princeton University Press 1997.

Patterson, William Patrick: 'Struggle of the Magicians – exploring the teacher-student relationship', Arete Communications, California 1998.

Peters, Fritz: 'Boyhood with Gurdjieff', Capra Press, Santa Barbara 1980.

Pirsig, Robert M.: 'Zen and the Art of Motorcycle Maintenance', Corgi Books 1976.

Sabians, Wikepedia: http://en.wikipedia.org/wiki/sabians. Internet June 2012.

Salaman, Clement; 'The Role of the Pagan Gods in Ficino', Temenos Academy papers 4, 2001.

Scholem, Gershom G.: 'Majors Trends in Jewish Mysticism', Schocken Books Inc. 1961.

Tarnas, Richard: 'The Role of Astrology in a Civilization in Crisis', The Astrological Association of Great Britain 2014.

Temenos Academy Review (selected papers from) 1981-2013.

Temple, Richard: 'Icons and the Mystical Origins of Christianity', Element Books Limited 1990.

The Gospel of Thomas, Marvin Meyer (translator), Harper Collins San Francisco 1992.

The Holy Bible, Authorised King James version, Collins Great Britain.

Thom, Alexander: 'Megalithic Sites in Britain', Oxford, Clarendon Press 1967.

Trevor-Roper, H.R.: 'The Persecution of Witches' in 'Light of the Past – a treasury of Horizon', American Heritage Publishing 1965.

Versluis, Arthur: 'Theo Sophia – hidden dimensions of Christianity', Lindisfarne Press 1994.

von Hofe, Hal: 'Seven Complete Hermetic Planetary Rituals Reconstructed from the Practices of the Sabians of Harran', Ra-Hoor Khuit Network's Magical Library. Internet July 2012.

Voss, Angela: 'God or Daemon? Platonic Astrology in a Christian Cosmos' in the Journal of the Temenos Academy 14, 2011.

Waterlow, Nick, and Mellick, Ross: 'Spirit and Place – art in Australia 1961-1996', Museum of Contemporary Art 1997.

Weburn, Andrew: 'The Beginnings of Christianity – Essene mystery, Gnostic revelation and the Christian vision', Floris Books 1995.

Wiseman, P.J.: 'Clues to Creation in Genesis', Marshall Morgan and Scott 1977.

Yates, Frances A.: 'Giordano Bruno and the Hermetic Tradition', University of Chicago Press 1964.

Yates, Frances A.: 'The Rosicrucian Enlightenment', Paladin 1975.

FURTHER ACKNOWLEDGEMENTS

My gratitude is expressed to the bush astrologer Mick Aboud, who presented me with 'Raphael's Astrology' when I was 20 years of age and who encouraged me; and to the late Alan Leo, whose simple descriptions of Rising Signs provided a shock of recognition when I needed it during those confusing younger years.

To Bonnie Matthews who taught a class of astrological comrades in Adelaide, where we learned from each other.

To George Ogilvie for being my first real client. To the many loyal clients, some of whom have become close friends; and to the four that became astrologers themselves.

To the countless astrology books consulted and to the master astrologers who took the time out to write them and so provide a real blessing for our tribe. My particular thanks go to Donna Cunningham, John Addey, Liz Greene, Robert Hand and Stephen Arroyo, none of whom I ever met but whose volumes still grace my old desk. You have paved the way for the rest of us to make our own discoveries.

ABOUT THE AUTHOR

Jane Ahlquist is an astrologer and spiritual teacher living in the Australian bush near the tiny town of Majors Creek in New South Wales. She has practised astrology for over 35 years. Jane pioneered practical weekends on women's spirituality in Sydney in the late 1980s and has adapted and presented traditional stories for theatre based upon her knowledge of astrological structures.

Jane gives consultations, in person and by post, and she regularly presents workshops and talks on astrology and related subjects.

For further information, please go to www.unicornhouse.org or contact Jane at artandspirit@antmail.com.au

www.ingramcontent.com/pod-product-compliance
Lightning Source LLC
Chambersburg PA
CBHW021035210326
41598CB00016B/1031